Modelling Crop-Weed
Interactions

Modelling Crop-Weed Interactions

Edited by

M.J. Kropff
International Rice Research Institute
PO Box 933
1099 Manila
The Philippines

and

H.H. van Laar
Department of Theoretical Production Ecology
Wageningen Agricultural University
PO Box 430
Bornsesteeg 65
6700 AK Wageningen
The Netherlands

CAB INTERNATIONAL

in association with the
International Rice Research Institute

CAB INTERNATIONAL
Wallingford
Oxon OX10 8DE
UK

Tel: Wallingford (0491) 832111
Telex: 847964 (COMAGG G)
Telecom Gold / Dialcom: 84: CAU001
Fax: (0491) 833508

Published in association with:

International Rice Research Institute
PO Box 933
1099 Manila
The Philippines

A catalogue record for this book is available from the British Library.

ISBN   0 85198 745 1 (CABI)
ISBN   971 22 0038 8 (IRRI)

Printed in Great Britain by BPCC Wheatons Ltd, Exeter

Contents

Contributors

Fajardo, F.F., International Rice Research Institute, P. O. Box 933, 1099 Manila, Philippines

Joenje, W., Department of Vegetation Science, Plant Ecology and Weed Science, Wageningen Agricultural University, Bornsesteeg 69, 6708 PD Wageningen, the Netherlands

Keulen, N.C. van, Department of Theoretical Production Ecology, Wageningen Agricultural University, P. O. Box 430, 6700 AK Wageningen, the Netherlands

Kraalingen, D.W.G. van, Department of Theoretical Production Ecology, Wageningen Agricultural University, P. O. Box 430, 6700 AK Wageningen, the Netherlands. [*]Centre for Agrobiological Research, P. O. Box 14, 6700 AA Wageningen, the Netherlands

Kropff, M.J., [*]International Rice Research Institute (IRRI), P. O. Box 933, 1099 Manila, Philippines. Department of Theoretical Production Ecology, Wageningen Agricultural University, P. O. Box 430, 6700 AK Wageningen, the Netherlands

Laar, H.H. van, Department of Theoretical Production Ecology, Wageningen Agricultural University, P. O. Box 430, 6700 AK Wageningen, the Netherlands

Lindquist, J.L., International Rice Research Institute, P. O. Box 933, 1099 Manila, Philippines

Lotz, L.A.P., Centre for Agrobiological Research, P. O. Box 14, 6700 AA Wageningen, the Netherlands

Migo, T.R., International Rice Research Institute, P. O. Box 933, 1099 Manila, Philippines

Schnieders, B.J., Department of Theoretical Production Ecology, Wageningen Agricultural University, P. O. Box 430, 6700 AK Wageningen, the Netherlands. [*]Centre for Agrobiological Research, P. O. Box 14, 6700 AA Wageningen, the Netherlands

[*] Current address

Foreword

We dedicate this book to the late Kees Spitters, who initiated much of the work that is described here. His work started with his PhD thesis in 1979, on the role of competition for selection in plant breeding. In the early 1980s, he began to develop more general descriptive models for competition than those that were then available, such as the replacement series models. Simultaneously, he initiated the development of mechanistic models for interplant competition. Together with his students, he developed the first eco-physiological competition models from the existing crop growth simulation models, which can be seen as an important breakthrough in understanding of phenomena related to interplant competition. He remained actively involved in the further development of the models until he died in 1990. Many of the ideas described here were developed during intense discussions with Kees and other colleagues and have been published in joint papers.

This book is the result of work conducted for more than a decade in collaborative research projects with many colleagues. The work started at the Department of Theoretical Production Ecology of the Wageningen Agricultural University, The Netherlands, in close collaboration with the Centre for Agrobiological Research (CABO-DLO) and continued at the International Rice Research Institute in the Philippines. It resulted in a number of approaches, ranging from relatively simple regression models for prediction purposes to detailed physiological models that can be functional research tools.

We felt that there was a need for a comprehensive description of the approaches that were developed to enable others to use them as transparent tools instead of black boxes. The book contains much information that can be useful to non-modellers as well. However, we hope that the models will find application in weed research as a tool, such as statistics is now. We hope that this book will bridge the gap between modellers and practical weed scientists, and that the

knowledge will help to improve integrated weed management systems.

Los Banos, Philippines   Martin J. Kropff
May 1993   H. H. van Laar

Preface

Competition between plants for the capture of the essential resources for plant growth (light, water, and nutrients) is a critical process in natural, semi-natural, and agricultural ecosystems. (Agro)ecologists have studied interplant competition intensively. However, because of the complex nature of interplant competition, it has been difficult to develop generalizing concepts and theories.

This book describes the most widely used available descriptive models that have been applied to describe competition effects in replacement and additive competition experiments. The potential of such simple models for predicting yield loss by weed competition is discussed. Attention is paid to eco-physiological understanding of competition between plants on the basis of the distribution of the resources light, water, and nutrients over species, and the way species utilize the amounts taken up for dry matter production. These mechanisms are discussed within the framework of a simulation model for interplant competition (INTERCOM). The focus is on the understanding of competition between weeds and crops, that is needed to improve weed management systems. The mechanisms, however, are universal, making the approach useful in other interplant competition situations as well.

A detailed description of the model INTERCOM and its assumptions is given, as well as a methodology to parameterize and evaluate the model using experimental data. It is written in such a way that all key equations are explained to enable the reader to understand the program listing, an essential requirement for using an eco-physiological simulation model. Examples are given of how the model can be used to understand the effects of weeds on crops in field situations. Several applications of the model in weed science are highlighted. How physiological and morphological characteristics of species or varieties are related to their competitive ability is shown. The ultimate application will be use of the model by scientists who have specific hypotheses related to competition they would like to pre-test. This could help improve the experiments conducted to provide the evidence.

The book can be used as a comprehensive guide for (weed) ecologists or intercropping specialists who want to obtain a better understanding of phenomena observed in their experiments by using an eco-physiological approach. It could also provide a framework for students in these disciplines to obtain more integrated, quantitative insight into the complexity of competition in plant communities. However, it should be noted that the model reflects current knowledge on understanding the processes underlying competition.

Acknowledgements

Many colleagues have been involved in the work presented in this book, and especially in developing the models described here and on conducting experiments to validate them. From the Department of Theoretical Production Ecology, Wageningen Agricultural University were involved Kees Spitters and Nel van Keulen. From the Centre for Agrobiological Research: Bert Lotz, Wim de Groot and Roel Groeneveld. From the Department of Vegetation Science, Plant Ecology and Weed Science, Wageningen Agricultural University: Gert Liefstingh and Wouter Joenje. From the International Rice Research Institute: Keith Moody, Ferdi Fajardo and Ted Migo.

Through an intensive collaborative project with Susan Weaver, Roger Cousens, Joy Rooney and John Porter on wild oat - wheat competition, other scientists contributed as well. The stay of Susan Weaver in Wageningen and Long Ashton, and the visits of Martin Kropff and Roger Cousens to Harrow, made possible by a NATO grant, have been instrumental.

Part of the work conducted in Wageningen was co-funded by an EC grant (EV4C-0020-NC)

The following students contributed to the work in various ways and were often involved in the generation of new ideas: Arnoud ten Cate, Gerhard Coster, Rien Aerts, Frans Vossen, Don Jansen, Gert-Jan Noij, Piet Spoorenberg, Frans Pladdet, Frits van Evert, Sanderine Nonhebel, Jan Geertsema, Winand Smeets, Lammert Bastiaans, Barbara Habekotté, Harmke van Oene, Rob Werner, Willem Hensums, Hans Peelen, Jaap Stroet, Ivonne Vlaswinkel, Marie-Antoinette Smit, Luc Mangnus, Marieke van Suylen, Sjaak Conijn, Els Verberne, Bert Schnieders, Gert-Jan van Delft, Marijke Vonk, Angela Jansen, Huib Hengsdijk, Gerard Keurentjes, Rob Stokkers, Peter Schippers, Bert Bos, Jacco Wallinga, and John Lindquist.

We thank C.T. de Wit for his many useful discussions on the topic we had in the past decade. Jens Streibig and Roger Cousens critically commented on the first draft of the manuscript and gave very useful suggestions to strengthen the chapter on applications.

We gratefully acknowledge the detailed and constructive comments on several versions of the manuscript of Keith Moody. His critical comments from a practical weed scientist's point of view helped us to keep the focus on practical weed management and its problems.

Say Calubiran-Badrina's invaluable help in making the figures is very much appreciated.

Abstract

An eco-physiological model is presented which simulates competition between plants from the distribution of the resources light, water and nutrients over the species and the way the species utilize the amounts taken up for dry matter production.

Competition for light is simulated on the basis of the leaf area of the competing species and its distribution over the height of the canopy. The absorbed radiation by the species in relation to plant height is calculated first. Using the CO_2 assimilation light response of individual leaves, the profile of CO_2 assimilation in the canopy is calculated. Integration over height and the day gives the daily rate of CO_2 assimilation of the species. After subtraction of losses for maintenance and growth, the daily growth rate in dry matter of the species is obtained. Effects of drought and nutrients are taken into account by a simple water and nutrient balance in which the available amounts of soil moisture and nutrients during the growing season are tracked. Soil moisture and nutrients are allocated to the competing species mainly proportional to their demands. The model is general and can be used for any plant species. A detailed description is given of the way in which values for model parameters can be derived from experimental data. The model was used to analyse several field experiments in different climatic environments. The model performed well for most situations. The model can be used as a tool to analyse the backgrounds of differences in competitive effects between treatments in experiments. The model was used to analyse the contribution of several species characteristics to the competitiveness of a species. Potentials for use of the described model in research and agricultural practice with specific attention to weed management are discussed.

Chapter One

General Introduction

M.J. Kropff

Competition: a key process in (agro-)ecology

Interplant competition for capture of the essential resources for plant growth (i.e. light, water and nutrients) is one of the key processes determining the performance of natural, semi-natural and agricultural ecosystems. Although farmers must have recognized competition effects in their systems as soon as they started to shape ecosystems to meet their needs, the first scientific reports on competition were published in the 14th century (Grace and Tilman, 1990). Darwin (1859) identified the central role of competition in selection processes of organisms in general. Since then, competition has been regarded as one of the major forces behind the appearance and life history of plants and the structure and dynamics of plant communities (Grace and Tilman, 1990).

Because of its important role in a wide range of ecosystems, competition between plants has been studied from different perspectives. In ecology, scientists have studied competition between plants to understand succession patterns of vegetation, the diversity and stability of plant communities and to help define management strategies for semi-natural ecosystems (Grace and Tilman, 1990; Grime, 1979; Harper, 1977). In agricultural sciences, competition studies focused on minimizing the effect of weeds or 'unwanted plants' by optimization of crop plant densities, and on development of predictive tools for yield loss assessment to develop weed management systems with minimum herbicide inputs (Zimdahl, 1980; Radosevich and Holt, 1984; Altieri and Liebman, 1988). Much research on competition between plants has been conducted to maximize the output of intercropping systems by optimizing sowing times and densities (Willey, 1979a, b; Vandermeer, 1989). In forestry, the main thrust has been the optimization of tree stand densities and thinning practices to maximize wood production. In an excellent series of reviews that were

1

compiled by Grace and Tilman (1990), the different perspectives on plant competition were discussed and analysed.

Many authors have developed definitions of competition, ranging from broad to very narrow. These definitions are generally linked to a specific theoretical framework developed for a specific type of system in which competition is studied. This has led to several controversies among scientists, the most recent one being the debate about the widely used theories of the ecologists Grime (1979) and Tilman (1988). Grime (1979) defined competition as 'the tendency of neighbouring plants to utilize the same quantum of light, ion or mineral nutrient, molecule of water, or volume of space'. The competitive ability of a species is then determined by the capacity to capture and exploit resources rapidly. Tilman (1987) defined competition as 'the utilization of shared resources in short supply by two or more species'. The competitive ability of a species is then determined by its minimum resource requirement called R^*. Grace (1990) concluded that the theories are not contradictory but complementary: if the habitat is fertile, the competitive ability of a species is determined by its resource capture capacity, whereas in low fertility situations the competitive ability is related to the capacity of a species to tolerate low resource availability.

In agricultural systems, crops are grown at moderate to high resource levels. In many of these systems large amounts of resources (nutrients, water) are added to the system to maximize yields. Competition in these systems could be defined as the process of capture and utilization of shared resources by the crop and its associated weeds. In the specific situation of crop-weed competition, dealt with in this book, focus is on the effect of resource capture by weeds on crop growth and production. Those resources of which the supply cannot meet the demand are of major interest, as they determine the attainable yield of the crop. If weeds capture such resources, crop growth will be reduced resulting in yield losses.

It is well recognized that the actual mechanisms of competition for resource capture by plants are not simple. Plants are morphologically and physiologically extremely plastic in their response to their environment, making generalization of plant responses difficult.

Competition and weed management

Worldwide a 10% loss of agricultural production can be attributed to the competitive effect of weeds, in spite of intensive control of weeds in most agricultural systems (Zimdahl, 1980). Without weed control, yield losses range from 10 - 100%, depending on the competitive ability of the crop (van Heemst, 1985). Therefore, weed management is one of the key elements of most agricultural systems. The use and application of herbicides was one of the main factors enabling intensification of agriculture in the past decades. However, increasing herbicide resistance in weeds, the necessity to reduce cost of inputs, and widespread concern about environmental side effects of herbicides, have resulted in great pressure on farmers to reduce the use of herbicides. This led to the development of strategies for integrated weed management based on the use of alternative methods for weed control and rationalization of herbicide use. Rather than trying to eradicate weeds from a field, emphasis is on the management of weed populations. It has been shown that weed control in some crops (like winter wheat) is generally not needed to reduce yield loss in the current crop, but only to avoid problems in future crops (Lotz *et al.*, 1990). The development of such weed management systems requires thorough quantitative insight in the behaviour of weeds in agroecosystems and their effects. This involves both insight in crop-weed interactions within the growing season as well as the dynamics of weed populations over growing seasons.

Several attempts have been made to develop weed control advisory systems, using thresholds for weed control, i.e. the level of weed infestation which can be tolerated based on specified criteria which are generally based on economics (cf. Niemann, 1986; Aarts and de Visser, 1985; Wahmhoff and Heitefuss, 1988). A number of concepts for thresholds for tactical (within season) and strategic (long-term) decision-making in weed management have been developed (Cousens, 1987). However, the approach has hardly been used in practice (Cousens, 1987; Norris, 1992). Besides problems related to accuracy in yield loss predictions, good quantitative data on the effects of specific weeds in specific crops are sparse as well as reliable simple assessment methodologies. These problems have resulted in major constraints to the development and implementation of weed control advisory systems (H.F.M. Aarts, Research Station for Arable Farming and Field Production of Vegetables, Lelystad; and B. Gerowitt, Institute for Crop Protection, Göttingen, personal communications).

Norris (1992) concluded based on an extensive survey that in spite of the vast literature on effects of weed density and duration of competition on crop losses, the information generated had hardly any impact on weed management. He strongly argued that weed science should focus more on the biology of the weeds and develop understanding of the mechanisms of competition to design management programs.

Methods to quantify plant competition

Studies that aimed at quantification of interplant competition started
around 1900. Plants were grown in monocultures and mixtures of
species or cultivars. It was found that plants growing in monoculture
yielded differently from plants growing in mixtures (Montgomery,
cited by Spitters, 1990). One of the first systematic approaches to
study competitive phenomena was developed by de Wit (1960). He
introduced an experimental design (the replacement series in which
the mixing ratio varies, but total density remains constant) together
with a mathematical model to analyse the results. Since then this
approach has been used intensively in agricultural and ecological sciences to study dynamics of populations, competitive ability of plants
and the advantage of intercropping systems (reviews by Radosevich
and Holt, 1984; Trenbath, 1976; Harper, 1977; Vandermeer, 1989).
Several papers have discussed disadvantages and pitfalls of the
approach like the dependence of the model coefficients on total stand
density (de Wit, 1960; Connolly, 1986; Taylor and Aarssen, 1989).

Several more specific mathematical models for agronomic purposes, such as the prediction of yield loss by weeds (added to the
standard crop density), were developed (review by Spitters, 1990).
However, only in the 1980s was an approach developed to describe
competition over a range of population densities with varying mixture
ratios and at a range of total densities (Suehiro and Ogawa, 1980;
Wright, 1981; Spitters, 1983a). Similar approaches have been developed using the neighbouring approach in which the number of plants
sharing the same space as the central plant is related to the weight of
the central plant (Silander and Pacala, 1985; Pacala and Silander,
1987; Firbank and Watkinson, 1987). These descriptive regression
models are based on the same principle as the approach of de Wit
(1960): the non-linear hyperbolic relationship between yield and plant
density. Cousens (1985) reviewed a wide range of mathematical expressions to describe crop yield losses by weeds, and concluded that
this hyperbolic yield loss - weed density relationship was superior to
other equations.

These regression models provide a simple and accurate description of the competition effects in a particular experiment in which
only weed density is varied. However, regression coefficients may
vary strongly among experiments due to factors other than weed
density like the period between crop and weed emergence (cf. Kropff,
1988a; Cousens *et al.*, 1987; Håkansson, 1983). Some researchers introduced an additional variable in the hyperbolic yield - density equation to account for the effect of differences in the period between crop

and weed emergence (Håkansson, 1983; Cousens *et al.*, 1987). However, in practice weeds often emerge in successive flushes, making it difficult to apply a descriptive model that accounts for the effect of both weed density and the relative time of weed emergence, because weeds of different flushes have to be counted separately. Another disadvantage of these descriptive models is that they do not give insight into the competition process itself. Analysis of the causes of variation in coefficients is, therefore, only possible by performing laborious and expensive empirical studies across sites and years.

Recently, an alternative approach was derived from the hyperbolic yield - density relationship which relates relative leaf area by the weeds to yield loss (Kropff and Spitters, 1991). This approach partly helps to overcome the limitations of the earlier mentioned regression approaches, because the relative leaf area accounts for the density as well as the relative age of the weeds. This approach has potential for practical application because of its simplicity.

However, all these approaches have a phenomenological character: the outcome of competition is described at a given moment in time, but no explanation is given of the process. The extrapolation domain of these descriptive approaches is often limited, because they only account for the effect of a small number of factors that influence the competition process. They also cannot help to identify the mechanisms for evolution of species traits or for the structure and dynamics of populations, communities and ecosystems (Tilman, 1988) and they cannot help explaining the mechanisms for variation in competition effects between seasons and sites (Kropff, 1988a).

Toward a mechanistic understanding

The complexity of relationships between morphological and physiological characteristics of species and competitive ability of these species in mixtures has been recognized by many researchers. However, quantitative studies which are focused on unravelling these relationships are rare. Several attempts have been made in the 1980s to develop more mechanistic approaches to interplant competition. In the early eighties, Spitters and colleagues started to develop eco-physiological models for interplant competition, based on eco-physiological models for monoculture crops (Spitters and Aerts, 1983; Kropff *et al.*, 1984). They focused on crop-weed competition for light and water. These models are based on the assumption that competition is a dynamic process, that can be understood from the distribution of the limiting resources between the competing neighbouring plants and the efficiency with which each plant uses the resources captured. So, weeds and crops are interacting by changing the environment and resource availability. Such a mechanistic approach provides insight into the processes underlying competition effects observed in (field) experiments and so may be of help in searching for ways to manipulate competitive relations, like those between crop and weeds.

Several others followed similar approaches to quantify crop-weed interactions (Graf *et al.*, 1990b; Wilkerson *et al.*, 1990) or competition between grassland species (Rimmington, 1984). In ecology, mechanistic models for interplant competition have been developed with focus on competition for nutrients and succession in more complex semi-natural plant communities (e.g. Tilman, 1988; Berendse *et al.*, 1987; Berendse *et al.*, 1989). Because they study long-term effects (large number of years instead of one growing season), their approach is strongly different from the approach taken in the crop-weed interaction studies where instantaneous processes are taken into account.

Overview

This book discusses the different quantitative approaches to crop-weed competition that have been developed, ranging from empirical regression models (Chapter 2) to a detailed eco-physiological systems approach (Chapters 3 - 10).

Chapter 2 describes the hyperbolic yield - density relationships and their use in monocultures, replacement series and addition series. An extended version of the model presented by Cousens *et al.* (1987) that also accounts for the effect of timing of weed emergence is discussed as well. Also, an alternative approach for yield loss prediction based on the relative leaf area of weeds is described in detail and potentials and limitations are discussed.

In Chapter 3, an overview is given of the process-based eco-physiological approach that was taken to develop the model INTERCOM (INTERplant COMpetition). This model is used as a framework for the discussion of competition mechanisms for the resources light (Chapter 4), water (Chapter 5) and nutrients (Chapter 6).

In Chapter 4, processes related to competition for the resource light are discussed. After a discussion of the phenomena involved, the mechanistic approach taken in the model INTERCOM is discussed in detail. Chapter 5 deals with competition for the resource water, and Chapter 6 with the resource nitrogen.

An extensive description of the process to derive quantitative data to describe the characteristics of the species at the process level is given in Chapter 7. The resulting parameter values for the model INTERCOM are required to simulate interplant competition using the model. Instead of characterizing the competitive ability of a species by a single number, the processes underlying competitive ability are quantified here by parameter values.

In Chapter 8, results of a range of studies are presented in which the model was used to analyse crop-weed competition effects in field situations. Special attention is paid to the process of unravelling the importance of the different factors determining competition for light, water and nitrogen in field situations.

Subsequently, the model was used to analyse the impact of environmental factors (climate, density, date of emergence) and genetic factors (characteristics related to morphological and physiological processes) on the outcome of competition, which forms the basis for breeding efforts to increase the competitive ability of crops.

Finally, several applications of the process-based model INTERCOM are presented in Chapter 10. This chapter specially focuses on how the model can be used as a research tool, and the complementary role of process-based eco-physiological models and simple regression models, which are more suitable for practical applications.

Chapter Two

Empirical Models for Crop-Weed Competition

M.J. Kropff and L.A.P. Lotz

Introduction

Many empirical models or regression equations have been developed to describe the effect of weeds on crop yield. Most of these models predict yield loss from the weed density. In this chapter, different forms of the most widely used hyperbolic equation for the relation between weed density and yield loss will be described. Also extensions of the approach and the recently developed relative leaf area model will be described.

Competition in monocultures

Agronomists have studied competition in monocultures extensively in the past to optimize harvestable yield and minimize seed inputs in the system by conducting density experiments. Montgomery (1912, cited by de Wit, 1960) already reported the crop yield - weed density response of oats, showing that yields increased with reduced spacing until a maximum yield level was reached. In the 1950s and 1960s, a group of Japanese scientists identified the major effects of competition in monocultures and developed approaches to describe relationships that appeared to be consistent mathematically (Kira *et al.*, 1953; Shinozaki and Kira, 1956; Yoda *et al.*, 1963). They related plant size and plant survival (self-thinning) to competition effects as a result of differences in plant spacing. In those years, de Wit (1960) started his analysis of competition in mixtures using the hyperbolic equation for the yield - density response as well.

The rectangular hyperbola is the most widely used approach to describe the density response (Fig. 2.1):

$$Y_{cm} = N_c / (b_0 + b_c N_c) \qquad (2.1a)$$

9

where
Y_{cm} is the yield of the crop in monoculture in g m^{-2},
N_c the plant density of the crop in numbers m^{-2},
b_0 the intercept, and
b_c is the slope.
The reciprocal per plant weight then is a linear function of plant density:

$$1/W_c = N_c / Y_{cm} = b_0 + b_c N_c \qquad\qquad (2.1b)$$

where W_c is the weight per plant (g plant^{-1}). The intercept b_0 is the reciprocal of the virtual biomass of an isolated plant. It deviates from the real biomass of an isolated plant, because at wide spacing, without interplant competition the biomass is density independent and does not decrease with density as the hyperbola suggests. This is illustrated in Fig. 2.1B where the relation between $1/W_c$ and density levels bends off at low densities. The parameter b_c is the reciprocal of the maximum biomass per unit area i.e. the asymptote in Fig. 2.1A.

     This hyperbola can be used to describe the response of total dry matter production of crops to density reasonably well except for low densities where plants are growing in isolation and do not compete with each other. For harvestable yield, however, a very narrow optimum density has been observed for several species like maize. In cereals, seed yield and total dry matter production are constant over a wide range of densities, in which the crop is able to form a closed canopy. If harvestable yield shows an optimum shape of function an alternative model has to be used (Spitters, 1983b).

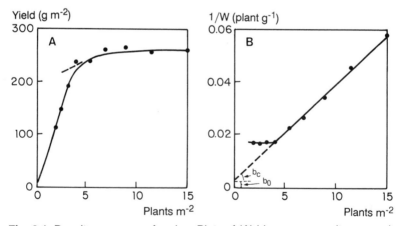

Fig. 2.1. Density response of maize. Plots of (A) biomass per unit area and (B) the reciprocal of per-plant weight versus plant density. Source: Spitters, 1983a.

In plant monocultures, interplant competition may result in mortality of the least competitive plants. This phenomenon is known as self-thinning, which was first described by Yoda *et al.* (1963). They found a close relationship between the number of plants per unit area that survived and the weight per plant. Their so-called 2/3 power law describes the dynamics of the system i.e. how density declines as mean weight per plant increases in a single stand in time. This relationship has been intensively used in forestry research, to determine thinning rates for tree stands, to optimize timber production. For a detailed discussion of these approaches we refer to Firbank and Watkinson (1990).

Hyperbolic functions for competition in mixtures

The analysis of competition in mixtures based on the hyperbolic yield density function for a monoculture was first introduced by de Wit (1960). He introduced a set of equations together with an experimental design: the replacement series approach. In this approach, the relative proportions of the components of a mixture are varied, but the total density is kept constant. This approach became widely used to determine the competitive ability of species, to address questions related to the real yield advantage of intercropping over monoculture cropping etc. The yield of a species is expressed relative to its yield in monoculture. The sum of the relative yields is the *RYT* (Relative Yield Total). If *RYT*>1, there is a true advantage of mixed cropping. In the replacement series design, some pitfalls have been identified (e.g. Connolly, 1986; Taylor and Aarssen, 1989). For example, the results are very sensitive to the total density of the mixtures (e.g. Firbank and Watkinson, 1985). In other words, the model coefficients vary with total density.

In weed science, replacing plants is not realistic. An additive approach is needed in which plants (weeds) are added to the standard crop density instead of a replacement approach (Fig. 2.2). Several mathematical equations have been used to relate crop yield loss to weed density (see review by Cousens, 1985).

An alternative approach was developed by several workers in the early 1980s (Suehiro and Ogawa, 1980; Wright, 1981; Spitters, 1983a). This approach is based on the same principles as the approach of de Wit (1960). The starting point in the derivation of the model is the response of crop yield to plant density, which can be described by a rectangular hyperbola (Eqn 2.1). Because $1/W_c$ is linearly affected by adding plants of the same species, it seems reasonable to assume that adding plants of another species also affects $1/W_c$ in a linear way (Spitters, 1983a):

$$Y_{cw} = N_c / (b_{c0} + b_{cc} N_c + b_{cw} N_w) \qquad (2.2a)$$

$$Y_{wc} = N_w / (b_{w0} + b_{ww} N_w + b_{wc} N_c) \qquad (2.2b)$$

The reciprocal per plant weight then equals:

$$1/W_c = b_{c0} + b_{cc} N_c + b_{cw} N_w \qquad (2.3a)$$

$$1/W_w = b_{w0} + b_{ww} N_w + b_{wc} N_c \qquad (2.3b)$$

where Y_{cw} and Y_{wc} are the yields of the crop and the weeds in a mixture respectively, and N_c and N_w are the number of crop and weed plants, respectively. The parameters b_{cc} and b_{ww} measure intraspecific competition between plants and the parameters b_{cw} and b_{wc} measure interspecific competition effects between the species. Fig. 2.3 demonstrates the shape of such a relationship for maize and *Echinochloa crus-galli* (Spitters *et al.*, 1989b). Adding 11 maize plants to a pure stand of *E. crus-galli* had the same effect on $1/W_c$ as adding 28 *E. crus-galli* plants. This approach allows the analysis over a range of total densities and mixing ratios. The parameter values can be used to derive indices for the relative competitive ability of the weeds (b_{cw}/b_{cc}) and niche differentiation: if the ratio $(b_{cc}/b_{cw})/(b_{wc}/b_{ww})$ exceeds unity, there is niche differentiation and $RYT>1$, indicating that the mixture as a total captures more resources than the respective monocultures. This can happen when species have a different rooting system, exploiting different compartments of the soil or in the situation of legumes, intercropped with other crops that can use the nitrogen fixed by Rhizobia.

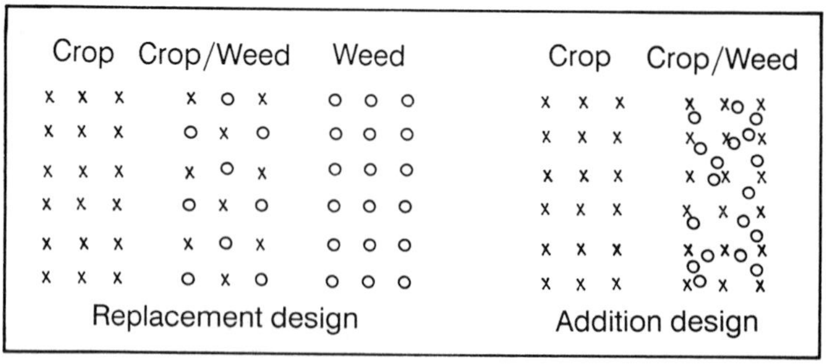

Fig. 2.2. Replacement and addition design with crop plants (x) and weed plants (o). Source: Spitters, 1990.

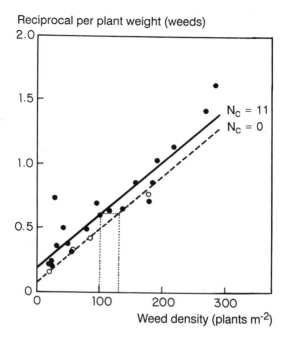

Fig. 2.3. Effect of *E. crus-galli* density on the reciprocal of the per-plant weight of *E. crus-galli* in plots without maize (o, N_c = 0) and in plots with maize (●, N_c = 11). The dotted lines indicate that addition of 11 maize plants to a population of 100 *E. crus-galli* plants m⁻² had the same effect as adding 28 *E. crus-galli* plants to that population. Regression equations were $1/W = 0.075 + 0.00416 \cdot N$ for $N_c = 0$, and $1/W = 0.196 + 0.00410 \cdot N$ for $N_c = 11$. Source: Spitters *et al.*, 1989b.

The difference between the approach of de Wit (1960) and the approach introduced by Spitters (1983a) is the result of several assumptions made by de Wit (1960), making it necessary to keep total density constant. However, if his Eqn 8.5 (de Wit, 1960, p. 60) is split up, so that the coefficients for both species do not have to be the same, the same set of equations is obtained (Spitters, 1990).

In general, the crop is grown at a fixed density, simplifying the situation to an additive series. Eqn 2.2a then simplifies to:

$$Y_{cw} = N_c / (a_0 + b_{cw} N_w) \text{ or } 1/W_{cw} = a_0 + b_{cw} N_w \qquad (2.4)$$

where $a_0 = b_{c0} + b_{cc} N_c$, which is equal to $1/W_{cc}$, the reciprocal of the average weight per plant in the weed free crop. The yield of a weedy crop can be expressed by

$$Y_{cw}/Y_{cm} = a_0 / (a_0 + b_{cw} N_w) = 1/ (1 + b_{cw} N_w /a_0) \qquad (2.5)$$

or for yield loss:

$$YL = 1 - (Y_{cw}/Y_{cm}) = 1 - (1/ (1 + b_{cw} N_w /a_0)) \qquad (2.6)$$

or

$$YL = aN_w / (1 + aN_w) \qquad (2.7)$$

where $a = b_{cw}/a_0$, describing the yield loss caused by adding the first weed m^{-2}; YL gives the relative yield loss, and N_w is the weed density. The yield reduction by a mixed weed population can be quantified by: $aN_w = a_1N_1 + ... + a_nN_n$ for weed species 1 to n. In this one-parameter regression model, maximum yield loss is 1 (or 100%) at high weed densities. The model fits experimental data accurately (Fig. 2.4).

The parameters can be estimated from the linearized form (Eqn 2.3) by linear regression. However, when plant weights are distributed normally, their reciprocals (1/W) show a skewed distribution and variances increase in plant density (Fig. 2.4B). A weighted regression should be applied in this case (Spitters *et al.*, 1989b). However, it is more convenient to use non-linear regression techniques on the logarithm of both sides of the equation because the yields are distributed log-normally (Fig. 2.4A) (Spitters *et al.*, 1989b).

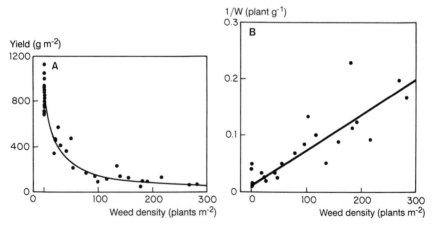

Fig. 2.4. Effect of *E. crus-galli* density on (A) maize biomass per unit area, and (B) reciprocal of per-plant weight of maize (1/W). Relations were described by the regression equation 1/W = 0.00132 + 0.00058 • N_{weeds} and Y_{crop} = 11.1 • W_{crop}. Source: Spitters *et al.*, 1989b.

Cousens (1985) introduced a hyperbolic yield loss - weed density equation which involves an additional parameter which permits a maximum yield loss of less than 100% (m):

$$YL = aN_w / (1 + aN_w/m) \qquad (2.8)$$

where
YL is the relative yield loss (%),
N_w the weed density (plants m^{-2}),
a parameter that describes the effect of adding the first weed (m^2 plant^{-1}), and
m is the maximum relative yield loss.

Cousens (1985) demonstrated the superiority of this equation over others by statistical analysis. This model has an advantage over Eqn 2.7 in situations where a clear maximum yield loss is observed (Fig. 2.5). Such a maximum yield loss can be explained biologically. For example in situations where the weeds cannot overtop the crop or when weeds emerge late, yield loss at high weed densities cannot approach 100%, as the crop is the stronger competitor.

Because variation in the yield loss - weed density function is often largely determined by differences in the period between crop and weed emergence (Cousens *et al.*, 1987; Håkansson, 1983; Kropff *et al.*, 1984; Kropff *et al.*, 1992a), precise prediction of yield loss on the basis of early observations should be based on both weed density and the period between crop and weed emergence. An additional variable in the hyperbolic yield loss - weed density equation to account for the effect of differences in the period between crop and weed emergence

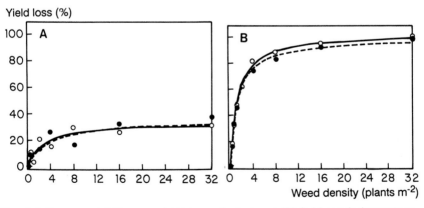

Fig. 2.5. Percent yield losses of (A) transplanted and (B) field-seeded tomatoes as a function of nightshade density at Harrow (Canada) in 1984 (o) and 1985 (●) and the fitted hyperbolic curves (Eqn 2.8) for 1984 (solid line) and 1985 (dashed line). Source: Weaver *et al.*, 1987.

was introduced by Håkansson (1983) and Cousens *et al.* (1987).

The regression model for the effect of both factors, weed density and the period between crop and weed emergence on yield loss, developed by Cousens *et al.* (1987) was mathematically formulated as:

$$YL = \frac{x \, N_w}{\exp(y \bullet T_{cw}) + (x/z)N_w} \tag{2.9}$$

in which
YL is the relative yield loss,
N_w is the weed density (plants m^{-2}),
T_{cw} is the period between crop and weed emergence (d),
x, y and z are non-linear regression coefficients.
This regression model fitted very well simulated data on yield loss at different densities and dates of weed emergence, although the regression model was unable to describe the effect of very early emerging weeds. A problem of this approach is the great need for data, because the effect of weed density has to be studied at a range of dates of weed emergence. A second problem in relation to practical application is that weeds often emerge in successive flushes. Every flush has a different competitive ability and weed densities of different flushes have to be distinguished. Therefore, alternative approaches are needed to predict yield loss by weeds for use in weed management systems.

The relative leaf area approach

The hyperbolic relative leaf area - yield loss equation

An alternative approach was suggested by Spitters and Aerts (1983) and Kropff (1988a). It was shown by simulation studies that a close relationship exists between yield loss and relative leaf area of the weeds, determined shortly after crop emergence (Kropff, 1988a; see Chapter 10). In these simulation experiments, both density of the weeds and the period between crop and weed emergence were varied over a wide range. Based on these findings, a new simple descriptive regression model for early prediction of crop losses by weed competition was developed by Kropff and Spitters (1991). This model was mathematically derived from the well-tested hyperbolic yield loss - weed density model and relates yield loss (YL) to relative weed leaf area (L_w expressed as weed leaf area / crop + weed leaf area) shortly after crop emergence, using the 'relative damage coefficient' q. This derivation is described in the following section.

A simple one parameter expression for yield loss (*YL*) as a function of the relative weed density (N_w/N_c) can be derived from Eqns 2.1 and 2.2 when crop density is constant:

$$YL = 1 - \frac{Y_{cw}}{Y_{cm}} = \frac{a\,\dfrac{N_w}{N_c}}{1 + a\,\dfrac{N_w}{N_c}} \qquad (2.10)$$

where *a* characterizes the competitive effect of the weed on the crop:

$$a = \frac{b_{cw}\,N_c}{b_{c0} + b_{cc}\,N_c} \qquad (2.11)$$

Generally, the crop is grown at such densities that the monoculture yield (Y_{cm}) approaches its maximum value, so that the parameter b_0 (Eqn 2.1) can be neglected. The expression for the parameter *a* then approaches:

$$a = \frac{b_{cw}}{b_{cc}} \qquad (2.12)$$

Because the competitive strength of a species is determined by its density and relative time of emergence, it is better to express the presence of the species by its *LAI* during early growth. This results in an expression for yield loss as a function of the ratio between the *LAI* of the weed and the *LAI* of the crop:

$$YL = 1 - \frac{Y_{cw}}{Y_{cm}} = \left(q\,\frac{LAI_w}{LAI_c}\right) \Big/ \left(1 + q\,\frac{LAI_w}{LAI_c}\right) \qquad (2.13)$$

with the single parameter *q* which is called 'the relative damage coefficient'. This relative damage coefficient is directly related to the parameter *a* in Eqns 2.10 and 2.11, because the *LAI* is the product of the leaf area per plant (*LA*) and the plant density:

$$LAI = N \bullet LA \qquad (2.14)$$

The relative damage coefficient *q* can then be expressed as (by combining Eqns 2.10, 2.13 and 2.14):

$$q = a\,\frac{LA_c}{LA_w} \qquad (2.15)$$

in which LA_c and LA_w are the average leaf areas per plant of the crop and the weed at the moment of observation. This can be interpreted as a weighting of the coefficient a, which only accounts for density effects, by the relative leaf area of the species to account for the period between crop and weed emergence as well. Eqn 2.15 shows that several small weed plants have the same effect as one older and bigger weed plant.

However, a parameter which may be easier to estimate than the ratio of the leaf area indices is the share in total leaf area of the weed species (L_w):

$$L_w = LAI_w / (LAI_w + LAI_c) \qquad (2.16)$$

or

$$LAI_w / LAI_c = 1 / ((1/L_w) - 1) \qquad (2.17)$$

The model in Eqn 2.13 can be reformulated to express yield loss of the crop as a function of the relative leaf area of the weeds:

$$YL = \frac{q\, L_w}{1 + (q - 1)\, L_w} \qquad (2.18)$$

The model in Eqn 2.18 can easily be extended in an additive way to allow for more weed species:

$$YL = \frac{\Sigma q_i\, L_{w,i}}{1 + \Sigma(q_i - 1)\, L_{w,i}} \qquad (2.19)$$

When the crop is grown at such a density that monoculture yield reaches its maximum value and the crop and weeds have identical physiological and morphological characteristics, the relative damage coefficient q approaches unity and a linear relation is obtained (the diagonal 1:1 line; Fig. 2.6). When the weed is a stronger competitor than the crop, the relative damage coefficient q will be larger than one and a convex curve is found above the diagonal line. When the crop is the stronger competitor, the relative damage coefficient q will be smaller than one and a concave curve is found under the diagonal line. The theoretical relations for different values of the relative damage coefficient q are shown in Fig. 2.6.

Time dependence of relative damage coefficient q

To enable precise decision-making in weed management, yield loss caused by the weeds has to be estimated at the time of decision-making. However, the weed density changes in time if weeds keep on emerging, and the ratio of the leaf area per plant of the crop and the weed (LA_c/LA_w) may change if the plants have different growth characteristics. That implies a time dependence of the relative

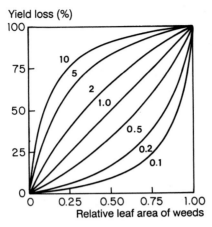

Fig. 2.6. Theoretical relations between yield loss and relative leaf area of the weeds according to Eqn 2.18 at different values of the parameter q. Source: Kropff and Spitters, 1991.

damage coefficient q. However, this time dependence can be quantified. In the early growth phase, when the observations on weed infestation have to be made, the canopy is not yet closed and the crop and weed plants generally grow exponentially according to the function:

$$LA_t = LA_0 \bullet e^{(R_l \bullet t)} \tag{2.20}$$

where

LA_t represents the leaf area per plant at time t,

LA_0 the leaf area at the reference time 0 (the moment of observation for which the relative damage coefficient q has been determined from experimental data),

R_l the relative growth rate of the leaf area ($^{\circ}C^{-1} d^{-1}$), and

t is the time expressed in degree days ($^{\circ}C$ d).

The relative growth rate of the leaf area R_l is only relevant in early growth phases when plants grow exponentially and can easily be determined by growth analysis of free growing plants.

From Eqns 2.20 and 2.15 it can be derived that the change in time of the relative damage coefficient q in the period of exponential growth when the canopy is not closed equals:

$$q = q_0 \bullet e^{(R_{l(c)} - R_{l(w)}) \bullet t} \tag{2.21}$$

where q_0 is the value of q when L_w is observed at $t = 0$ (the moment of observation for which the relative damage coefficient q has been determined from experimental data) and t indicates the period between $t = 0$ and the moment of observation (in degree days) for which the relative damage coefficient q will characterize the effects. When the weeds and the crop have the same value of R_l, the relative damage coefficient q will be constant in time for all different dates of observation. When q_0 is determined for a given crop weed combination at a certain period after crop emergence, the value of the relative damage coefficient q at other dates of observations can be calculated using Eqn 2.20.

Recently, another version of the model was derived from the empirical model introduced by Cousens (1985). This model includes an extra parameter for the maximum yield loss caused by weeds (m), and was derived in a similar way as Eqn 2.18 (Kropff and Lotz, 1992):

$$YL = \frac{q\,L_w}{1+(\frac{q}{m}-1)\,L_w} \tag{2.22}$$

or for a multi-species situation:

$$YL = \frac{\Sigma q_i L_{w,i}}{1+\Sigma(\frac{q_i}{m_i}-1)\,L_{w,i}} \tag{2.23}$$

Evaluation of the relative leaf area approach

The relative leaf area - yield loss model was evaluated using data on the effect of *Monochoria vaginalis* (Burm. f.) Presl and *Echinochloa crus-galli* L. (Beauv.) on transplanted rice (*Oryza sativa* L.); *Chenopodium album* L., *Stellaria media* L. and *Polygonum persicaria* L. on sugar beet (*Beta vulgaris* L.) and *E. crus-galli* L. on maize (Kropff and Spitters, 1991) and *Solanum pthycanthum* Dun. in transplanted tomatoes (Kropff *et al.*, 1992a). The approach appeared to be superior to the yield loss - weed density relationship because it accounted for the effect of density and the period between crop and weed emergence. Two examples will be given here.

In transplanted rice, early emerging weeds (five days after transplanting) reduced yield by 21% (*M. vaginalis*) up to 26% (*E. crus-galli*) at the highest weed densities. However, late emerging weeds hardly reduced crop yield. This indicates that weed density is not an effective measure for weed infestation for prediction purposes (Fig. 2.7). The relationship between the relative leaf area index of *M. vaginalis* and

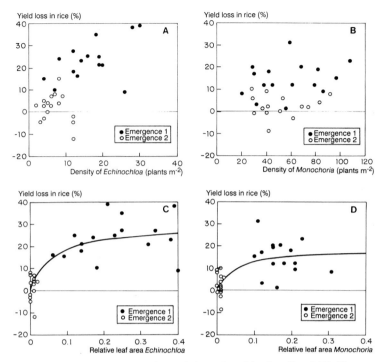

Fig. 2.7. Yield loss in transplanted rice related to density of (A) *E. crus-galli* and (B) *M. vaginalis* and relative leaf area index of (C) *E. crus-galli* and (D) *M. vaginalis* determined 36 days after transplanting. Observed values: (●) early emergence weeds (five days after transplanting), (o) late emerging weeds (22 days after transplanting), and the solid line is fitted with Eqn 2.18.

E. crus-galli and yield loss in rice resulted in a better description of the data. Both the relative damage coefficient q and the maximum yield loss were smaller for *M. vaginalis*. This can be explained by the strong difference in plant height between the species: rice was able to overtop *M. vaginalis* but not *E. crus-galli*.

As a second example, the effects of *C. album* and *P. persicaria* on sugar beet yield are presented in Figs 2.8 and 2.9. Fig. 2.9 shows that the yield loss - weed density relation strongly differs between weeds that emerged at different periods after the crop, but that a reasonable relationship was found when yield loss was related to the relative leaf area of the weeds shortly after crop emergence. However, if curves are fitted for a range of observation dates for the different weed flushes separately, a trend in the relative damage coefficients is observed:

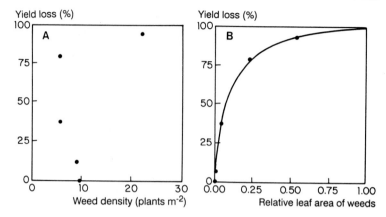

Fig. 2.8. (A) Relationship between weed density and yield loss for five field experiments with sugar beet and *C. album*; (B) relationship between relative leaf cover of the weeds 30 days after sowing and yield loss for five experiments with sugar beet and *C. album*. Source: Kropff and Lotz, 1992.

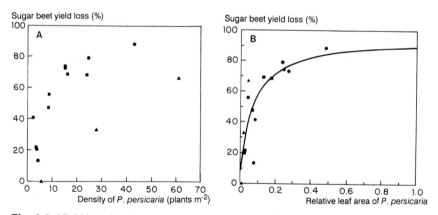

Fig. 2.9. Yield loss in sugar beet related to density (A) and relative leaf area index (B) *of P. persicaria*, determined 23 days after crop emergence in 1990. Observed values: weeds sown 5 (●), 10 (■) and 15 days after crop emergence (▲); solid line represents fit with Eqn 2.18.

later emerging weeds have a higher damage coefficient at a specific date of observation and the relative damage coefficient is smaller at later observation dates. Preliminary analyses showed that these effects can be roughly explained by the differences in relative growth rate using Eqn 2.21. However, more detailed studies have to be conducted to confirm those observations. So, a more tight relationship is found by the relative leaf area approach compared to the density approach, but for practical application, simple approaches to account for the time dependence have to be developed.

The complication of the time dependence of the relative damage coefficient q is a direct result of including the effect of the period between crop and weed emergence in the simple regression model by characterizing the weed infestation by its relative leaf area.

The relative leaf area model was also used by Ranganathan (1993) to estimate the impact of the perennial pigeonpea on intercropped groundnut plants. She concluded based on the analysis that in a system where pigeonpea is pruned before sowing the groundnut, yield loss is relatively low, but without pruning, yield loss can be considerable.

Evaluation of the model for multi-species competition

The ability of the model to describe the effect of mixed weed populations on crop yield was analysed by comparing measured yield loss to predicted yield loss using the q and m values determined for the individual weed species. The problem with Eqn 2.23 could be that the regression model does not account for interspecific competition between the weed species, which will affect their competitive ability. Fig. 2.10 shows the relation between predicted and observed sugar beet yield loss due to the combined effect of *S. media* and *C. album*. Since *C. album* was a much stronger competitor than *S. media*, the trend in calculated yield loss is mainly determined by *C. album*. A tendency to underestimate the effect of late emerging weeds and to overestimate the effect of early emerging weeds was observed (Fig. 2.10). This may be the result of the slightly different effects of early and late emerging *C. album* on maximum yield loss.

Conclusions

A wide range of regression models have been developed to describe competition effects. Most approaches relate yield to plant density. The addition approach (e.g. Spitters, 1983a) can be viewed as an extended version of the replacement series. These approaches are very helpful for analysis of competition experiments with a range of densities of the species. The most widely used hyperbolic yield loss - weed density equation which was introduced by Cousens (1985) describes experimental data accurately. However, alternative approaches will be needed for yield loss predictions, because the yield loss - weed

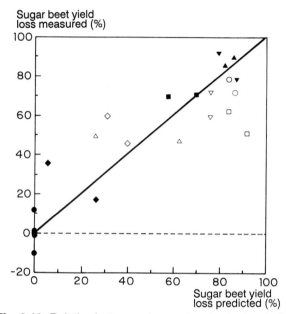

Fig. 2.10. Relation between observed and calculated with Eqn 2.23, using parameters from single weed species treatments determined by Eqn 2.22, yield loss in a mixture of sugar beet with *C. album* and *S. media*; emergence of weeds 5 days after crop emergence (DAE): □ (1), o (2), ▼ (3); weed emergence 10 DAE △ (1), ∇ (2), ▲ (3); weed emergence 15 DAE: ◆ (1), ◊ (2), ■ (3), ● (control). (Numbers indicate densities (1) 1.4 *C. album* m^{-2} and 1.4 *S. media* m^{-2}; (2) 2.8 *C. album* m^{-2} and 5.6 *S. media* m^{-2}; (3) 5.6 *C. album* m^{-2} and 44 *S. media* m^{-2}).

density relation varies among situations because of differences in the period between crop and weed emergence.

Such an approach could be the relative leaf cover - yield loss model that accounts for the effect of weed densities, the period between crop and weed emergence and different flushes of the weeds, within a limited range of time after crop emergence. However, it should be noted that the effect of other factors, such as transplanting shock or severe water stress, is not accounted for (Kropff and Spitters, 1991). An aspect that needs further attention is the dependence of q on the time of observation after crop emergence when the relative growth rate of the leaf area of crop and weeds differs (Eqn 2.21) (Kropff and Spitters, 1991).

Chapter Three

Eco-Physiological Models for Crop-Weed Competition

M.J. Kropff

Introduction

Simulation models for the behaviour of natural and agricultural ecosystems have been developed and used in the past three decades. They have been extremely helpful in integrating knowledge from various disciplines in one framework and have helped to improve insight into complex ecosystems. The approach is characterized by the terms systems (a limited part of the real world), models (a simple representation of a system) and simulation (constructing mathematical models and analysing the behaviour of the model in relation to the behaviour of the real system). Generally the so-called state variable approach is chosen, based on the assumption that the state of the system can be quantified at any moment in time, and that changes in the state can be described by mathematical equations (de Wit and Goudriaan, 1978). This leads to models in which state variables, rate variables and driving variables are distinguished.

State variables are quantities that can be measured, such as biomass, numbers, amount of water in the soil. Each state variable is associated with rate variables that characterize the rate of change at a given moment in time as a result of a specific process such as photosynthesis or respiration. These rate variables represent flows of material between state variables. Their value depends on the state variables and driving variables based on knowledge of the processes in the system and not on statistical behaviour. Driving variables characterize the effect of the environment of the system at its boundaries, for example, meteorological variables, such as temperature, rain, radiation. The models based on these principles attempt to explain the behaviour of the system based on the underlying processes. For further background information on the philosophy behind the approach we refer to de Wit (1982) and Penning de Vries (1982).

An approach to crop simulation modelling

The starting point of the model described here is the approach used in modelling the growth of monoculture crops. These models simulate the growth of crops based on the response of eco-physiological processes to the environment. A very helpful classification system for crop production systems was introduced by de Wit and Penning de Vries (1982). They distinguished four production situations:

Production Situation 1 The potential production situation, where water and nutrients are available to the crop in ample supply. Crop growth and production are only determined by radiation, temperature and species characteristics. Models for this production situation are the core of any model for crop growth.

Production Situation 2 In this production situation, crop growth is limited by water supply for at least a part of the growing season. A water balance of the soil has to be included in the model.

Production Situation 3 In the third production situation, water and nitrogen are limiting crop growth and production for at least part of the growing season. This is the situation in many rainfed agro-ecosystems.

Production Situation 4 In the fourth production situation, phosphorus or other nutrients limit crop production. This is the complex situation in systems where no fertilizers are being used.

In all these production situations, the effect of growth reducing factors like pests, diseases and weeds can be included.

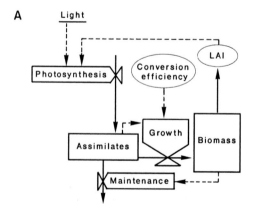

Fig. 3.1. (A) Relational diagram of a system in Production Situation 1, where radiation and temperature are the growth determining factors (continued).

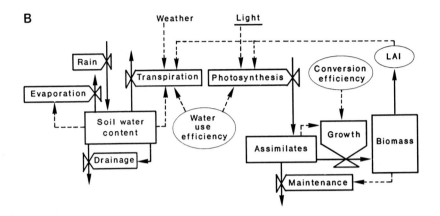

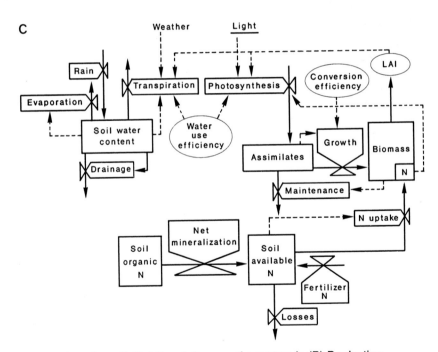

Fig. 3.1 (continued). Relational diagram of a system in (B) Production Situation 2 , where water is limiting crop growth at least part of the growing season; (C) Production Situation 3, where nitrogen availability limits growth of the crop. Boxes represent state variables, valves represent rate variables, circles represent intermediate variables, lines are flows of material, and broken lines are flows of information.

Models for the first two production situations have been developed in the past three decades and have been thoroughly evaluated by now (van Keulen, 1975; de Wit *et al.*, 1978; Rabbinge *et al.*, 1989; Penning de Vries *et al.*, 1989; Jones and Kiniry, 1986; Jones *et al.*, 1989; Kropff *et al.*, 1993b). Models for the third production situation have been developed, but simulation of soil-N dynamics and N uptake by the crop is still in its preliminary stage, because of the complexity of soil processes (van Keulen and Seligman, 1987; de Willigen, 1991). Models for the fourth production situation are generally simple and not eco-physiological, because of the lack of understanding of the processes involved (Janssen *et al.*, 1987; Janssen *et al.*, 1990; Wolf *et al.*, 1987). This classification system enables a systematic approach in modelling. The core of any eco-physiological crop growth model is the potential production model for Production Situation 1. Models for the other production situations are basically extensions of that model (Fig. 3.1).

The approach taken here is to develop a model with a broad applicability, but focused on a specific set of problems. The objective is not only to simulate the system satisfactorily, but mainly to use the model to help solving practical problems in relationship to crop-weed management. For example, to use the model to help defining how the system can be manipulated to reduce problems related to intensive use of herbicides and other weed control measures, or to define which plant characteristics are important in relation to competitiveness of the crop for breeders, or to develop predictive tools for yield loss by weeds to avoid unnecessary use of herbicides. However, the first objective was to develop a model to obtain quantitative understanding of the system. The first question to be addressed was why weeds cause more problems in one situation compared to other situations. For example, a tremendous difference was found in the effect of *Echinochloa crus-galli* on maize yield in two consecutive seasons at the same site (Kropff *et al.*, 1984). The complexity of the system requires tools like simulation models to obtain understanding of the behaviour of the system.

Another important aspect of the approach taken, is the attempt to make the model as simple as possible. That enables the modeller to understand the behaviour of the model itself. Complexity is only added if the performance of the model in relation to the real system requires improvement. In the process of development of the model for interplant competition, procedures to simulate some physiological processes have been simplified, whereas simulation procedures for other processes, that appeared to be critical for quantification of competition like leaf area development and height development, needed refinement. For a long time it has been recognized that simulation procedures for processes related to morphology were by far the weakest feature of the crop simulation models (de Wit and Penning de Vries, 1985). But until it was necessary to improve the simulation procedures for morphological traits in competition models, not much attention was paid to these processes. Some detailed models have been developed for leaf area development of monoculture crops but not for plant height (Jones and Kiniry, 1986; Muchow and Carberry, 1989; Porter, 1984). However, these morphological submodels may not be suitable for competition models, because a feedback between growth and morphological processes is needed in competition situations and the morphological plasticity of species has to be accounted for.

The model INTERCOM for interplant competition

Objectives

The main objective of the model INTERCOM is to provide a tool to analyse the complex interactions between plants that compete for the resources light, water and nitrogen. Focus is especially on crop-weed interactions and understanding of the backgrounds of differences in effects of different weed species on crops and the variation of effects in different environments. The model was designed to account for effects of temperature, radiation, rainfall and soil hydrological characteristics on plant growth in competition situations. The model has been used to analyse several competition situations in different agro-ecological environments. The effect of N was not included in the analysis of experimental studies, as N was always in ample supply in the experiments. However, a first approach to add the effect of N to the model is presented in this book (Chapter 6).

Approach and structure

The model simulates the following aspects of growth in competition:
- competition for capture of light and water,
- phenological development,
- morphological development (height and leaf area),
- dry matter accumulation,
- allocation of dry matter over the plant organs, and
- soil moisture balance and evaporation and transpiration.

Input requirements of the model are: geographical latitude, standard daily weather data (daily solar radiation, temperature, rainfall, average wind speed, vapour pressure), soil physical properties (wilting point and field capacity) and parameter values that describe the morphological and physiological characteristics of the plant species. In potential production situations, the water balance part of the model can be removed. Time step of integration is one day.

Model structure

The general structure of the modelling approach for monocultures is given in Fig. 3.2, and for mixtures in Fig. 3.3. Under favourable growth conditions light is the main factor determining the growth rate of the crop and its associated weeds. From the leaf area indices (*LAI*) of the species and the vertical distribution of their leaf area, the light profile within the canopy is calculated. Based on photosynthesis characteristics of single leaves, the photosynthesis profile of each species in the mixed canopy is obtained. Integration over the height of the canopy and over the day gives the daily CO_2 assimilation rate for each species. This is converted to glucose production by multiplying by 30/44 (molecular weights of CH_2O/CO_2). After subtraction of respiration requirements for maintenance of the species, the net daily

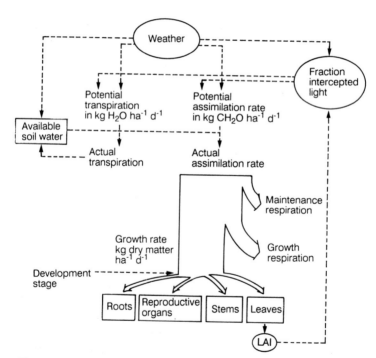

Fig. 3.2. General structure of the eco-physiological model for a monoculture crop.

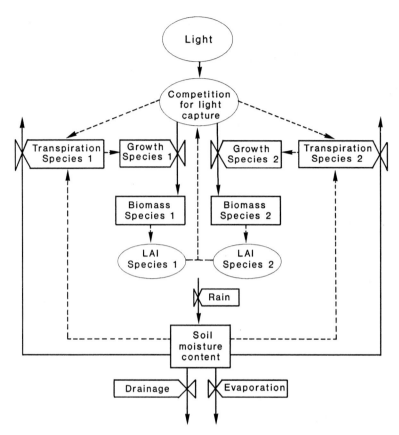

Fig. 3.3. General structure of the eco-physiological model for interplant competition (INTERCOM).

growth rate in kg dry matter per ha per day is obtained using a conversion factor. The dry matter produced is partitioned among the various plant organs, using partitioning coefficients that are introduced as a function of the phenological development stage of the species. Phenological development rate is tracked in the model as a function of ambient daily average temperature. When the canopy is not yet closed, leaf area increment is calculated from daily average temperature. When the canopy closes, the increase in leaf area is obtained from the increase in leaf weight using the specific leaf area (*SLA*, m^2 leaf kg^{-1} leaf). Integration of daily growth rates of the organs and leaf area results in the time course of *LAI* and dry weight in the growing season. This part of the model is described in Chapter 4.

Chapter 5 describes the additional modules needed to account for soil moisture stress effects. To account for the effects of drought stress, a simple water balance for a free draining soil profile is attached to the model, tracking the available amount of soil moisture in due time. Transpiration and growth rates of the species are reduced when the available soil moisture drops below a certain level. Competition for water is thus closely linked to above-ground competition for light, because transpiration is driven by the absorbed amount of radiation by the species and the vapour pressure deficit inside the canopy. Details of the model are presented in Chapter 4 (Production Situation 1), Chapter 5 (Production Situation 2) and Chapter 6 (Production Situation 3).

Chapter Four

Mechanisms of Competition for Light

M.J. Kropff

Introduction

In a potential production situation, where light, temperature and physiological and morphological characteristics determine the growth of a plant community, plants only compete for the resource light. In agricultural systems where other factors like nitrogen or water limit crop production, weeds compete with the crop for light as well as for the other resources. It is extremely difficult to distinguish the contributions of competition for the different resources to the total effect in an experimental way. A modelling approach helps to unravel the complex interrelationships. In this chapter, the process of competition for light is discussed only. Subsequent chapters deal with competition for other resources. The integration of these processes will be presented in Chapter 8, where the model INTERCOM is evaluated.

The availability of the resource light varies strongly during a day. Its intensity is relatively predictable following diurnal and seasonal schemes, although changes occur very rapidly due to clouds. The resource light cannot be stored as such in the system. Therefore, competition for light differs from competition for resources like water or nutrients. It is an instantaneous process of resource capture and the efficiency of resource capture is related to light interception and light use characteristics of the species. Understanding of the impact of factors involved may help in weed management and in breeding programs to develop competitive plant types.

Light interception in mixed canopies is determined by the leaf area index of the species, plant height, and light absorption characteristics of the leaves. If a leaf is positioned above another leaf it will absorb a larger amount of radiation. A strong correlation between plant height and competitive ability has been demonstrated for many crop species (reviewed by Berkowitz, 1988). However, there are often trade offs. In rice, for example, it has indeed been shown that taller varieties are more competitive (cf. Jennings and Aquino, 1967), but these taller varieties have a lower yield potential as a result of lodging and/or a reduced harvest index.

The light absorption characteristics of a species are related to leaf thickness and leaf angle distribution. The leaf angle distribution determines the amount of radiation absorbed per unit of leaf area. Planophile leaves (horizontally oriented) capture light with a higher efficiency than erectophile leaves (vertically oriented) (de Wit, 1965).

Besides these factors that determine the instantaneous process of light capture, the dynamics of the system during the growing period has to be taken into account and especially the feedback mechanisms involved. Plants that grow fast in early growth stages, often have a strong advantage and can build up a larger share in the canopy. The same holds of course when the competing species have different dates of emergence. For weeds, it has been well documented that the period between crop and weed emergence, which determines the relative starting position, strongly affects competitive effects of the weeds (Håkansson, 1983; Kropff, 1988a). Another important feedback mechanism is related to the reduction in final height of plants that emerge late and grow under a closed canopy. However, some species demonstrate a strong phenotypic plasticity with respect to height development. *Chenopodium album* L., for example, can overtop a sugar beet crop even when the weeds emerge under a closing canopy. Morgan and Smith (1981) demonstrated the strong impact of light quality on stem extension in *C. album*. They found that the decrease in the red-to-far-red ratio as a result of preferential absorption of red light by plants drives stem extension processes.

Phenological development of the species is important as well for the dynamics of competition. If one species matures earlier, the late maturing species may be able to recover when the leaves of the early species senesce. This aspect is often used in agricultural systems like crops growing under 'cover crops', that start growing after the dominant crop is harvested. This concept is used in relay cropping in which light use is optimized by partially overlapping cropping periods in intercropping systems (Willey, 1979a, b).

Besides characteristics related to light absorption, competitiveness has also been related to light use efficiency. Photosynthetic characteristics were reported to be highly correlated to the competitive ability of C_3 and C_4 species (Pearcy et al., 1981). However, because of the many factors involved, direct conclusions from correlations between competitive ability and a specific factor cannot be drawn without insight in the mechanisms.

Measuring of light capture by different species in a mixed canopy to improve understanding of the mechanisms is difficult. In situations where one species grows in the shade of another species, the resource availability for the shaded species can be measured just above its canopy. However, when the species form a mixed canopy the vertical profiles of leaf area and light intensity can be measured, but are difficult to interpret.

Modelling the light interception process is regarded as the most promising approach to understand light capture by species in mixed canopies (Berkowitz, 1988). Because many factors have to be considered in the evaluation of processes determining competition for light, a model with enough detail is needed to address questions related to the role of photosynthetic or morphological characteristics in light competition. Such models have been developed and tested in the past decade (cf. Spitters and Aerts, 1983; Kropff, 1988a; Kropff and Spitters, 1992; Ryel *et al.*, 1990; Graf *et al.*, 1990b). Mechanisms of competition for light are discussed in this chapter by using the framework of the detailed mechanistic modelling approach of the model INTERCOM. Simpler approaches, however, are described as well.

The approach taken in the model INTERCOM to simulate the growth of a mixture of species in a potential production situation is very similar to the approaches used in simulation of the growth of monoculture crops. The single difference is the extension of the calculation procedure for light absorption.

After a short overview of models that have been developed, the principles for simulation of light absorption in the canopy by competing species and the resulting rates of CO_2 assimilation of the species will be discussed. Subsequently, the procedure for simulation of growth and maintenance respiration is described. From CO_2 assimilation and respiration the total daily growth of the species is calculated. An alternative approach to simulate dry matter growth of the crop is described as well. In the following sections, procedures to simulate phenological development, dry matter partitioning and morphological development will be described. Procedures to estimate values for species parameters and functions that describe physiological and morphological characteristics from experimental data are described in Chapter 8.

Modelling competition for light

During the past decade, several models at different levels of detail have been developed to understand and predict competition for light capture. Spitters and Aerts (1983) introduced a detailed model of competition for light and water, based on existing crop growth models. Their model separated the leaf canopy into a large number of leaf layers. Light absorption and leaf photosynthesis were calculated by species for each leaf layer. Light absorbed in a layer was distributed according to the share in leaf area. Kropff *et al.* (1984) introduced an improved procedure by which the share in effective leaf area (the leaf area weighted by the extinction coefficient) was used instead of leaf area. The model has since been further improved and refined and evaluated with extensive data sets from field experiments (Kropff, 1988a; Spitters, 1989; Kropff and Spitters, 1992; Kropff *et al.*, 1992a, b; Weaver *et al.*, 1992). Van Gerwen *et al.* (1987) developed a similar model for tree canopy situations, using a procedure to account for horizontal heterogeneity of leaf area in the canopy.

A simpler approach was introduced by Rimmington (1984), who developed a model based on calculations of light absorption for a few leaf strata. It was indicated that care is needed to obtain the correct procedure for the different strata (Rimmington, 1984). Wilkerson *et al.* (1990) also introduced a simpler approach, defining a competition factor that indicates the efficiency of light absorption of the weeds relative to the crop. This factor is determined by calibrating the model with experimental data from mixtures. They do not explicitly account for height differences. Graf *et al.* (1990b) developed a model that distinguishes a number of layers equal to the number of competing species, with the heights of the species as the boundaries. Light absorption is calculated per layer and accumulated over the canopy. The model does not account for diurnal variation in radiation nor does it account for profiles of photosynthesis in the canopy.

A detailed model that only focused on competition for light was developed by Ryel *et al.* (1990), Beyschlag *et al.* (1990) and Barnes *et al.* (1990). They studied light absorption by species in mixtures of wheat and wild oats and used the model to determine the importance of photosynthetic characteristics and canopy structure in competitive situations. Takayanagi and Kusanagi (1991) also developed such a detailed model to study light competition in soybean - *Digitaria ciliaris* mixtures. The detailed approaches are very similar to the approach used in the model INTERCOM.

CO_2 assimilation of species in a mixed canopy

Radiation fluxes above the canopy

Measured or estimated daily total solar irradiation (wavelength 300 - 3000 nm) is input for the model. Only half of this incoming radiation is photosynthetically active (*PAR*, Photosynthetically Active Radiation, wavelength 400 - 700 nm). This fraction, generally called 'light', is used in the calculation procedure of the CO_2 assimilation rates of the species. A distinction is made between diffuse skylight, with incidence under various angles, and direct sunlight with an angle of incidence equal to the solar angle. It is important to distinguish these fluxes because of the large difference in illumination intensity between shaded leaves (receiving only diffuse radiation) and sunlit leaves (receiving both direct and diffuse radiation) and because of the non-linear CO_2 assimilation-light response of single leaves. The diffuse flux is the result of the scattering of sun rays by clouds, aerosols and gases in the atmosphere. The proportion of diffuse light in the total incident light flux depends on the status of the atmosphere, i.e. cloudiness, concentration of aerosols. This fraction is calculated from the atmospheric transmission using an empirical function. The atmospheric transmission is the ratio between actual daily total solar irradiance (measured $S_{g,d}$, J m^{-2} d^{-1}) and the quantity that would have reached the earth's surface in the absence of an atmosphere $S_{0,d}$ (J m^{-2} d^{-1}): $S_{g,d}$ / $S_{0,d}$. Daily values are taken here, as generally only daily totals are being measured at agro-meteostations.

The theoretical radiation flux (S_0) at the earth surface, assuming 100% atmospheric transmission, can be calculated from the solar constant (S_{sc}), which is the radiation flux perpendicular to the sun rays, and the sine of the solar elevation (β):

$$S_0 = S_{sc} \bullet \sin \beta \qquad (4.1)$$

So, S_0 follows a sinusoidal pattern over the day. The daily total theoretical radiation flux ($S_{0,d}$) is calculated by integrating S_0 over the day:

$$S_{0,d} = S_{sc} \bullet \int_{\text{sunrise}}^{\text{sunset}} \sin \beta \qquad (4.2)$$

Input requirements are the latitude of the site, the day of the year and the time of the day. A detailed description of these calculations is given by Spitters *et al.* (1986).

An empirically determined relationship is used in the model that relates $S_{g,d}$ / $S_{0,d}$ to the fraction of diffuse radiation ($S_{g,df,d}$ / $S_{g,d}$). This relationship is based on data from different meteorological stations from a wide range of latitudes and longitudes (Spitters *et al.*, 1986). The relationship is graphically represented in Fig. 4.1.

The radiation intensity strongly changes over the day, following the sine of the solar angle with the horizon (sin β). The relationship described by Spitters *et al.* (1986, their Eqn 6) is used to calculate the instantaneous flux densities of diffuse and direct light over the day. The atmospheric transmission is assumed constant over the day:

$$S_g / S_0 = S_{g,d} / S_{0,d} \qquad (4.3)$$

These instantaneous fluxes of diffuse and direct radiation are calculated in the Subroutines ASTRO and RADIAT (see Appendix 2). ASTRO calculates day length, declination of the sun, and intermediate variables, which are also used in RADIAT, based on day number and latitude using empirical functions. RADIAT calculates the instantaneous fluxes of diffuse and direct light at the specified moment of the day. The calculation procedure is illustrated in Fig. 4.1. The following sections deal with light profiles in a mixed canopy, CO_2 assimilation rates and the procedure to simulate total daily rates of canopy CO_2 assimilation from instantaneous CO_2 assimilation rates at different heights in the canopy. These procedures are programmed in the subroutines TOTASS and ASSIMC (Appendix 2).

Light profile within the mixed canopy

Incoming radiation is partly reflected by the canopy. The reflection coefficient (ρ) of a green leaf canopy with a random spherical leaf angle distribution, which indicates the fraction of the downward radiation flux that is reflected by the whole canopy can be approximated by (Goudriaan, cited by Spitters, 1986):

$$\rho = [(1 - \sqrt{(1 - \sigma)}) / (1 + \sqrt{(1 - \sigma)})] \bullet [2 / (1 + 1.6 \sin \beta)] \quad (4.4)$$

in which σ represents the scattering coefficient of single leaves for visible radiation ($\sigma = 0.2$; an average value for most crop species). The first term denotes the refection of a canopy with horizontal leaves (Goudriaan, 1977) and the second term is an approximate correction factor for a spherical leaf angle distribution (Goudriaan, personal communication). A fraction $(1 - \rho)$ of the incoming visible radiation can be absorbed by the canopy.

Radiation fluxes attenuate exponentially within a canopy with the cumulative *LAI*, counted from the top downwards:

$$I_L = (1 - \rho) I_0 \exp(-k L) \quad (4.5)$$

where
I_L is the net *PAR* flux at depth L in the canopy (the height in the canopy above which *LAI* equals L) (J m^{-2} ground s^{-1}),
I_0 is the flux of visible radiation at the top of the canopy (J m^{-2} ground s^{-1}),
L the cumulative leaf area index (counted from the top of the canopy downwards) (m^2 leaf m^{-2} ground),
ρ the reflection coefficient of the canopy (-), and
k the extinction coefficient for *PAR* (-).

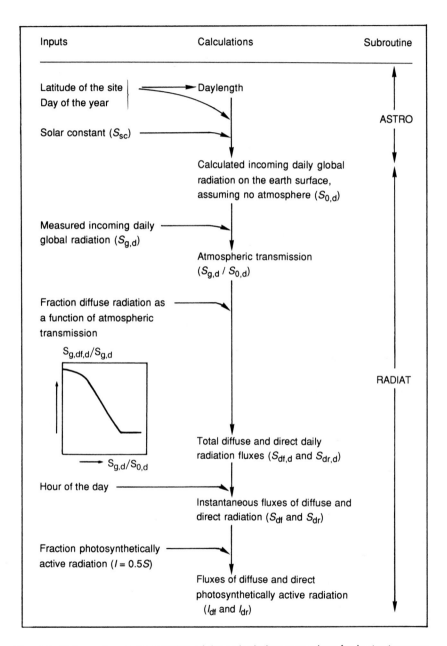

Fig. 4.1. Schematic representation of the calculation procedure for instantaneous fluxes of direct and diffuse solar radiation above the canopy from measured daily total global radiation.

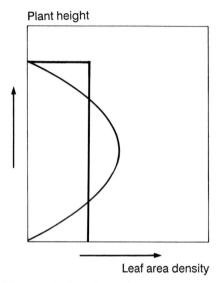

Plant height

Leaf area density

Fig. 4.2. Options for leaf area distribution over height of plants; the model: parabolic or constant.

The extinction coefficient k indicates the efficiency at which the foliage absorbs light, and is a function of leaf inclination and other factors, which will be discussed later.

This procedure (Eqn 4.5) works well for monoculture crops but not for mixtures of species with different heights. Therefore, in mixed canopies, the depth L should be expressed as a height h (m):

$$I_h = (1 - \rho)\, I_0 \exp(-k\, L_h) \tag{4.6}$$

where

L_h is the leaf area index above height h (m^2 leaf m^{-2} ground).

The leaf area above height h for a given species is calculated on the basis of the relationship between the leaf area density (LAD_h, m^2 leaf m^{-2} ground m^{-1} height) and the plant height. Two options are currently available in the model: (i) a constant leaf area distribution over height or (ii) a parabolic leaf area distribution, which is more realistic. The leaf area distribution functions are described in the Subroutines LEAFRE and LEAFPA, respectively. Both patterns are illustrated in Fig. 4.2. The constant leaf area density function (LAD_h) can be described by:

$$LAD_h = LAI\,/\,h_t \tag{4.7}$$

where

h_t represents total plant height (m), and

LAI   the total leaf area index of the species (m^2 leaf m^{-2} ground).

For a simple parabolic function with its maximum at $0.5 \bullet h_t$, the following relationship is used (for other leaf area density profiles in the canopy different relationships are needed):

$$LAD_h = (6 \; LAI \, / \, h_t^3 \,) \, (h_t - h) \bullet h \qquad (4.8)$$

For L_h, which is the cumulative leaf area index counted from the top of the plants down to height h, the following equation can be derived for a homogeneous leaf area distribution:

$$L_h = LAI \, (h_t - h) \, / \, h_t \qquad (4.9)$$

and for a parabolic relationship:

$$L_h = LAI \; - ((LAI \; / \, h_t^3) \, h^2 \, (3h_t - 2h)) \qquad (4.10)$$

In these equations, it is assumed that the leaves are homogeneously distributed over the total height of the plants. However, sometimes the leaf area distribution over plant height is skewed. Other functions can be included if data are available (Graf *et al.*, 1990b).

In a canopy with a mixture of species, Eqn 4.6 can now be rewritten as:

$$I_h = (1 - \rho) \, I_0 \, \exp(-\Sigma k_j \, L_{h,\,j}) \qquad (4.11)$$

where

I_h is the net flux at height h (J m^{-2} ground s^{-1}),

I_0 is the flux at the top of the canopy (J m^{-2} ground s^{-1}),

$L_{h,\,j}$ the cumulative leaf area index of species j above height h (m^2 leaf m^{-2} ground),

ρ the reflection coefficient of the canopy (-), and

k_j the extinction coefficient of species j (-).

The leaf areas ($L_{h,\,j}$), weighted by the extinction coefficients (k_j), are summed over the $j = 1,...,n$ plant species in the mixed vegetation.

The light absorbed by species i at a height h in the canopy ($I_{a,h,i}$, J m^{-2} leaf s^{-1}) is obtained by taking the derivative of Eqn 4.11 with respect to the cumulative leaf area index:

$$I_{a,h,i} = - \, dI_{h,i} \, / \, dL_i = k_i \, (1 - \rho) \, I_0 \, \exp(-\Sigma k_j \, L_{h,\,j}) \qquad (4.12)$$

The same equation is used to calculate the absorbed radiation by stems and reproductive organs, i.e. these organs are regarded as different species. The diffuse and the direct fluxes have different extinction coefficients, giving rise to different light profiles within the canopy for diffuse and direct radiation.

Therefore, three different radiation fluxes are distinguished: (*i*) the direct component of direct light (with extinction coefficient $k_{dr,bl}$, with *bl* for black since direct radiation becomes diffuse as soon as the sun ray is partly absorbed by a leaf), (*ii*) the total direct flux (with extinction coefficient $k_{dr,t}$) and (*iii*) the diffuse flux (with extinction coefficient k_{df}).

The extinction coefficient for the direct component ($k_{dr,bl}$) can be calculated as (Goudriaan, 1977):

$$k_{dr,bl} = 0.5 \, k_{df} / (0.8 \, \sqrt{(1 - \sigma)} \, \sin \beta) \qquad (4.13)$$

where

σ is the scattering coefficient for single leaves (-).

The extinction coefficient for the total direct flux ($k_{dr,t}$) can be calculated as (Goudriaan, 1977):

$$k_{dr,t} = k_{dr,bl} \, \sqrt{(1 - \sigma)} \qquad (4.14)$$

k_{df} is the measured extinction coefficient under diffuse sky conditions being input in the model. Other leaf angle distributions can be accounted for by the procedure described by Goudriaan (1986), which calculates k_{df} on the basis of the frequency distribution of leaves with angles in three classes (0 - 30°, 30 - 60° and 60 - 90°).

The absorbed fluxes for the different components per unit leaf area at height *h* in the canopy are:

$$I_{a,df,i} = - \, dI_{df,h,i} / \, dL_i = k_{df,i} \, (1 - \rho) \, I_{0,df} \, \exp(-\Sigma \, k_{df,j} \, L_{h,j}) \quad (4.15)$$

$$I_{a,dr,t,i} = - \, dI_{dr,t,h,i} / \, dL_i = k_{dr,t,i} \, (1 - \rho) \, I_{0,dr} \, \exp(-\Sigma \, k_{dr,t,j} \, L_{h,j})$$
$$(4.16)$$

$$I_{a,dr,dr,i} = - \, dI_{dr,dr,h,i} / \, dL_i = k_{dr,dr,i} \, (1 - \rho) \, I_{0,dr} \, \exp(-\Sigma \, k_{dr,dr,j} \, L_{h,j})$$
$$(4.17)$$

where

$I_{a,df,i}$ is the absorbed flux of diffuse radiation (J m^{-2} leaf s^{-1}),

$I_{a,dr,t,i}$ the absorbed flux of total direct radiation (J m^{-2} leaf s^{-1}), and

$I_{a,dr,dr,i}$ the absorbed flux of the direct component of direct radiation (J m^{-2} leaf s^{-1}).

The total absorbed flux for shaded leaves (J m^{-2} leaf s^{-1}) equals:

$$I_{a,sh} = I_{a,df} + (I_{a,dr,t} - I_{a,dr,dr}) \qquad (4.18)$$

For sunlit leaves the situation is more complicated. They absorb the flux that shaded leaves absorb as well as the direct component of the direct flux. However, the direct flux intensity differs for leaves with different orientation. The amount of the direct component of the direct flux absorbed by leaves perpendicular to the radiation beams equals:

$$I_{a,dr,dr} = (1 - \sigma) I_{0,dr} / \sin \beta \qquad (4.19)$$

$I_{a,dr,dr}$ is the direct component of incoming *PAR*. The amount of absorbed direct radiation by leaves depends on the sine of incidence at the leaf surfaces. Therefore, for sunlit leaves, CO_2 assimilation rates have to be calculated separately for leaves with different angles and have to be integrated over all leaf angles. This is described in the next section.

CO_2 assimilation rates of single leaves

The CO_2 assimilation-light response of individual leaves follows a saturation type of function, characterized by the initial slope (the initial light use efficiency, ε) and the asymptote (A_m) and is described by a negative exponential function (Goudriaan, 1982):

$$A_h = A_m (1 - \exp(-\varepsilon I_a / A_m)) \qquad (4.20)$$

where

A_h is the gross assimilation rate (kg CO_2 ha^{-1} leaf h^{-1}),

A_m the gross assimilation rate at light saturation (kg CO_2 ha^{-1} leaf h^{-1}),

ε the initial light use efficiency (kg CO_2 ha^{-1} leaf h^{-1} / (J m^{-2} leaf s^{-1})), and

I_a is the amount of absorbed radiation (J m^{-2} leaf s^{-1}).

The parameter ε is not strongly species dependent (Ehleringer and Pearcy, 1983), but A_m is. Both parameters are temperature dependent. In some of the plant specific routines A_m is a function of the leaf nitrogen content, e.g. rice.

From the absorbed light intensity at height h for one of the species, the assimilation rate of species i at that specific canopy height, can be calculated with Eqn 4.20. This procedure is followed for sunlit and shaded leaves separately.

For sunlit leaves, assuming a spherical leaf angle distribution, the rate of CO_2 assimilation is calculated by integration of the rate of CO_2 assimilation over the sine of incidence of the direct beam, using the Gaussian integration method (Goudriaan, 1986; see section on daily canopy CO_2 assimilation). The assimilation rate of species i per unit leaf area at a specific height in the canopy is the sum of the assimilation rates of sunlit and shaded leaves, taking into account the proportion of sunlit and shaded leaf area at that height in the canopy. This proportion is the fraction sunlit leaf area (f_{sl}) which equals the fraction of the direct radiation reaching that layer:

$$f_{sl} = \exp\left(-\Sigma\, k_{dr,bl} \bullet L_h\right) \tag{4.21}$$

where

$k_{dr,bl}$ is the extinction coefficient for the direct component of direct radiation, and

L_h the leaf area index above height h (including stem and flower areas!).

Instantaneous canopy CO_2 assimilation rates of the species

Canopy CO_2 assimilation of the species can be calculated by dividing the canopy in numerous small layers using numerical integration. However, to save computing time, it is calculated as the weighted average of the CO_2 assimilation rates at five selected heights h in the canopy using the Gaussian integration method (Goudriaan, 1986). This method specifies discrete points at which the value of the function to be integrated has to be calculated. It also defines the weighting factors that have to be applied to these values to obtain an accurate approximation compared to the analytical solution. The 5-point Gaussian integration procedure was used instead of the 3-point integration which is generally used in monoculture crops, because the radiation flux profile in the canopy is generally not simply exponential in mixed canopies with plants of different heights.

The procedure, used in the model INTERCOM, to simulate radiation fluxes and instantaneous rates of CO_2 assimilation in a mixed canopy is illustrated in Fig. 4.3.

In Figs 4.4 and 4.5, the results are given of the calculation procedure for a constant and a parabolic leaf angle distribution, respectively. Profiles of absorbed radiation, CO_2 assimilation and the cumulative CO_2 assimilation were calculated for leaf layers of 1 cm thickness instead of using the 5-point Gaussian integration method. In the mixtures, two species with the same characteristics and the same leaf area index ($LAI = 2$) but a different height are competing for light. The results reflect the effects on a cloudy day with 300 J m^{-2} s^{-1} as incoming diffuse light. The results show that simulations in which the rectangular leaf area distribution is used, give a reasonable estimation of canopy CO_2 assimilation in a mixed canopy in comparison to the more realistic situations in which leaves are distributed according to a parabolic shape.

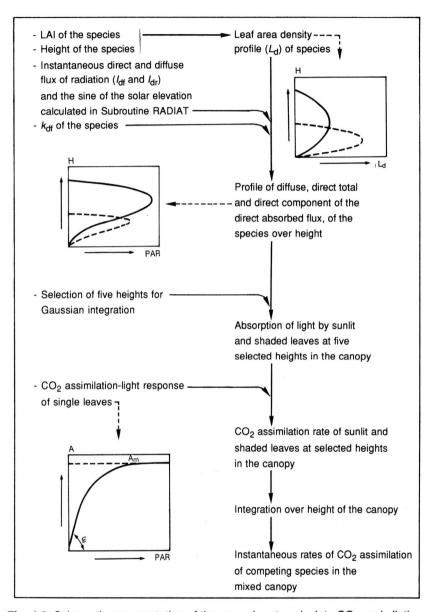

Fig. 4.3. Schematic representation of the procedure to calculate CO_2 assimilation of a canopy consisting of more than one species (Subroutines TOTASS and ASSIMC, see Appendix 2).

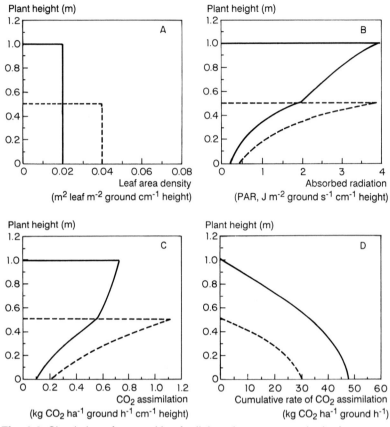

Fig. 4.4. Simulation of competition for light using a rectangular leaf area distribution. (A) Leaf area distribution over height for two identical species with both *LAI* = 2 and plant heights at 1 m (solid line) and 0.5 m (broken line). (B) Profile of absorbed visible radiation (*PAR*) for both species in mixture with 300 J m^{-2} s^{-1} diffuse radiation as input. (C) CO$_2$ assimilation profile in the canopy for both species (kg CO$_2$ ha^{-1} h^{-1} cm^{-1} height). (D) The cumulative CO$_2$ assimilation rate counted from the top of the canopy (kg CO$_2$ ha^{-1} ground h^{-1}).

The daily total canopy CO$_2$ assimilation rate of the species

The total daily rate of CO$_2$ assimilation of the species (A_d, kg CO$_2$ ha^{-1} ground d^{-1}) is obtained by integrating the instantaneous rates of CO$_2$ assimilation (A_h) over the height of the species and over the day. The integration is also achieved by applying the Gaussian algorithm. For

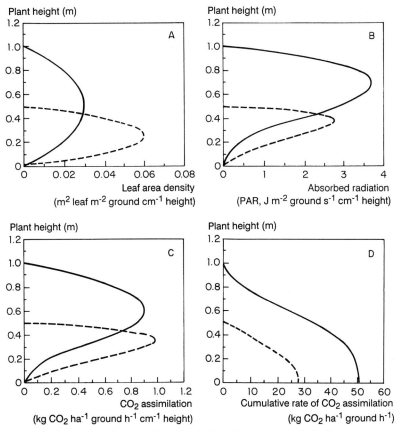

Fig. 4.5. Simulation of competition for light using a parabolic leaf area distribution. (A) Leaf area distribution over height for two identical species with both *LAI* = 2 and plant height of 1 m (solid line) and 0.5 m (broken line). (B) Profile of absorbed visible radiation (*PAR*) for both species in mixture with 300 J m^{-2} s^{-1} diffuse radiation as input. (C) CO_2 assimilation profile in the canopy for both species (kg CO_2 ha^{-1} h^{-1} cm^{-1} height). (D) The cumulative CO_2 assimilation rate counted from the top of the canopy (kg CO_2 ha^{-1} ground h^{-1}).

integration of the instantaneous assimilation rates over the height of the species in a mixture the 5-point method is generally satisfactory, while the 3-point method, which is less accurate, performs very well for integration over the day (Goudriaan, 1986). Thus, at three selected times during the day, the assimilation rates at five selected heights in the canopy are calculated for each species. The first version of the procedure, which did not include the light absorption by flowers and stems, was developed by S.E. Weaver, C. Rappoldt and M.J. Kropff at the Department of Theoretical Production Ecology, Wageningen Agricultural University. In the model INTERCOM, the absorption of light by flowers and stems is included.

In the program, the daily integration is executed in Subroutine TOTASS, and instantaneous rates of assimilation are calculated in the Subroutine ASSIMC (see Appendix 2).

Light absorption and CO_2 assimilation by stems and reproductive organs

In most models for canopy CO_2 assimilation, only light absorption by leaves is accounted for, although stems and reproductive organs like panicles absorb a substantial amount of radiation. In rice, for example, Stem (or sheath) Area Index (*SAI*) may go up to 1.5 m^2 stem m^{-2} ground and the Flower (panicle) Area Index (*FAI*, m^2 flower m^{-2} ground) may go up to 0.9 (M.J. Kropff and K.G. Cassman, IRRI, unpublished data). In the model, this is accounted for explicitly. For these organs, a homogeneous distribution of the area from the top to the bottom of the canopy is assumed, as it is for leaves. However, a subroutine based on a constant flower area density function, which accounts for the location of the area in the canopy, is available to make simulations more accurate or for studies that focus on analysing the impact of panicle position in the canopy (Kropff *et al.*, 1993a). Eqn 4.7 has to be changed into:

$$FAD_h = FAI / (h_t - h_0); \quad h > h_0 \qquad (4.22)$$

in which
FAD   is the flower area density (m^2 flower m^{-2} ground m^{-1} height),
FAI   the flower area index (m^2 flower m^{-2} ground), and
h_0 the lower boundary of the flower area (m).

Eqn 4.9 must also be changed into:

$$F_h = FAI (h_t - h) / (h_t - h_0); \quad h > h_0 \qquad (4.23)$$

The position of the organ in the crop has to be defined for that purpose. For rice, it was shown that a panicle in the top 10 cm of the canopy reduces canopy CO_2 assimilation by 25% (realistic situation), whereas a homogeneous distribution of panicle area over crop height would reduce canopy CO_2 assimilation by only 13% (Kropff *et al.*, 1993a).

Application of the mixed canopy model

The performance of the approach used in the model INTERCOM has been evaluated for crops in monoculture, using detailed experimental data on the diurnal course of canopy CO_2 assimilation of a *Vicia faba* L. crop on cloudy and clear days (Kropff, 1989). The model simulated instantaneous rates of canopy CO_2 assimilation very accurately. Measuring canopy CO_2 assimilation of species in a mixed canopy is not feasible, but as the model approaches the process very mechanistically, it can be used to study the impact of species characteristics. Ryel *et al.* (1990) developed a similar model and validated the model using measurements of fractions sunlit and shaded leaf area and light interception. The INTERCOM routines discussed above were used to analyse the effect of *LAI* and plant height on canopy CO_2 assimilation of two species. Fig. 4.6 shows that a lower *LAI* (factor 2) of a species can be compensated by a taller stature (factor 1.5), indicating the importance of relative height.

Maintenance and growth respiration

The assimilated CO_2 is converted into carbohydrates (CH_2O) in the CO_2 assimilation process. The energy for this reduction process is provided by the absorbed light. The overall chemical reaction of this complex process is:

$$6\ CO_2 + 6\ H_2O \xrightarrow{\text{light}} C_6H_{12}O_6 + 6\ O_2$$

or in a simplified form:

$$CO_2 + H_2O \xrightarrow{\text{light}} CH_2O + O_2$$

From this reaction, it follows that for every kg of CO_2 taken up, 30/44 kg of CH_2O is formed; the numerical values representing the molecular weights of CH_2O and CO_2, respectively. Part of the carbohydrates produced in this process are respired to provide the energy for maintaining the existing biostructures. This process is characterized in the model as maintenance respiration. The remaining carbohydrates are converted into structural plant dry matter. The losses in weight as a result of this conversion are characterized as growth respiration.

CO$_2$ assimilation (kg CO$_2$ ha^{-1} h^{-1})

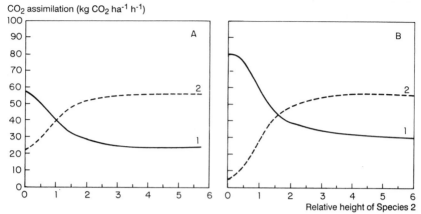

Fig. 4.6. The effect of *LAI* and plant height on CO$_2$ assimilation of a canopy consisting of two competing species. (A) Canopy CO$_2$ assimilation of the two species, Species 1 (solid line) and Species 2 (broken line) both *LAI* = 2, in relation to height of Species 2 relative to Species 1. (B) Canopy CO$_2$ assimilation of Species 1 (*LAI* = 4, solid line) and Species 2 (*LAI* = 2, broken line) in relation to height of Species 2 relative to Species 1.

Maintenance respiration

Maintenance respiration provides the energy for living organisms to maintain their biochemical and physiological status. Through the reaction which is the reverse of CO$_2$ reduction in CO$_2$ assimilation, the radiation energy which was fixed in the photosynthetic process in a chemical form is released in a suitable form (ATP and NADPH):

$$CH_2O + O_2 \text{ ---> } CO_2 + H_2O + \text{energy}$$

This process consumes roughly 15 - 30% of the carbohydrates produced by a crop in a growing season (Penning de Vries *et al.*, 1989), which indicates the importance of accurate quantification of this process in the model. However, the process is poorly understood at the biochemical level and simple empirical approaches are inaccurate since it is impossible to measure maintenance respiration in the way it is defined (Penning de Vries *et al.*, 1989; Amthor, 1984). The best way to quantify maintenance respiration is to measure the CO$_2$ production rate of plant tissue in the dark. The approach taken in the competition model is based on theoretical considerations, empirical studies, and studies in which the carbon balance in the model was evaluated using crop growth and canopy CO$_2$ assimilation data.

Three components of maintenance respiration can be distinguished at the cellular level: maintenance of concentration differences of ions across membranes, maintenance of proteins and a component related to the metabolic activity of the tissue (Penning de Vries, 1975). Maintenance respiration can thus be estimated from mineral and protein concentrations and metabolic activity as presented by de Wit *et al.* (1978). In the model INTERCOM, we use an adapted version of the simple approach developed by Penning de Vries and van Laar (1982a) based on the detailed model, in which maintenance requirements are approximately proportional to the dry weights of the plant organs to be maintained:

$$R_{m,r} = mc_{lv} W_{lv} + mc_{st} W_{st} + mc_{rt} W_{rt} + mc_{so} (1 - f_s) W_{so} \quad (4.24)$$

where for each species

$R_{m,r}$ is the maintenance respiration rate at the reference temperature (25 °C) in kg CH_2O $ha^{-1} d^{-1}$,

W_{lv}, W_{st}, W_{rt} and W_{so} are the weights of the leaves, stems, roots and storage organs (kg dry matter ha^{-1}), respectively;

mc_{lv}, mc_{st}, mc_{rt} and mc_{so} are the maintenance coefficients for leaves, stems, roots and storage organs, respectively; and

f_s is the fraction of material that is stored in the storage organs which have no maintenance requirements.

The maintenance coefficients (kg CH_2O kg^{-1} dry matter d^{-1}) have different values for the different organs because of large differences in nitrogen contents. Standard values for maintenance coefficients are 0.03 for leaves, 0.015 for stems and 0.01 for roots (Spitters *et al.*, 1989a). For tropical crops, like rice, lower values are used: 0.02 for leaves and 0.01 for other plant organs (Penning de Vries *et al.*, 1989).

The effect of temperature on maintenance respiration is simulated assuming a Q_{10} of 2 (Penning de Vries *et al.*, 1989):

$$R_m = R_{m,r} \bullet 2^{(T_{av} - T_r)/10} \quad (4.25)$$

where

T_{av} is the average daily temperature (°C), and

T_r is the reference temperature.

To account for the effect of metabolic activity, a special reduction factor is introduced in the model. This factor accounts for the reduction in metabolic activity and thus maintenance respiration when the crop ages. When nitrogen content is simulated in the model, this factor can be related to the N content (van Keulen and Seligman, 1987). In the current model, the total rate of maintenance respiration is assumed to be proportional to the fraction of the leaves that have been formed that is still green. This procedure for calculating the effect of age on maintenance respiration was used in the SUCROS model (Spitters *et al.*, 1989a) and was based on studies in which measured crop growth and canopy CO_2 assimilation data were analysed using a simple simulation model (Louwerse *et al.*, 1990; C.J.T. Spitters, CABO, unpublished data).

Growth respiration

The carbohydrates in excess of the maintenance costs are available for conversion into structural plant material. In the process of conversion, CO_2 and H_2O are released as scraps from the cut and paste process in biosynthesis. Following the reactions in the biochemical pathways of the synthesis of dry matter compounds (carbohydrates, lipids, proteins, organic acids and lignin from glucose (CH_2O)), Penning de Vries *et al.* (1974) derived the assimilate requirements for the different compounds. From the composition of the dry matter, the assimilate requirements for the formation of new tissue can be calculated. Typical values for leaves, stems, roots and storage organs have been presented by Penning de Vries *et al.* (1989).

Daily growth rate from CO_2 assimilation and respiration rate

The daily potential growth rate of the species (G_p, kg dry matter ha^{-1} d^{-1}) is calculated by:

$$G_p = (A_d \bullet (30/44) - R_m) / Q \qquad (4.26)$$

where

A_d is the daily rate of gross CO_2 assimilation of the crop (kg CO_2 ha^{-1} d^{-1}),

R_m the maintenance respiration costs (kg CH_2O ha^{-1} d^{-1}), and

Q the assimilate requirement for dry matter production (kg CH_2O kg^{-1} dry matter).

If R_m exceeds A_d, the net growth is set to 0. The value of Q is calculated as an average of the assimilate requirements of the different organs weighted with the fraction allocated to these organs.

A simple procedure for simulation of dry matter growth rates in mixed canopies

The above described procedure for simulation of plant growth in mixed canopies is included in the model INTERCOM. It contains much detail and can be simplified if one is not interested in specific processes. A simplified procedure was presented by Spitters (1989) and will be briefly described here. The performance is compared to the results of the INTERCOM procedure.

Dry matter growth rates of field crops appear to be more or less proportional to the amount of light intercepted (review by Gosse *et al.*, 1986). In the simplified version of the competition model, the detailed computations of CO_2 assimilation and respiration are replaced by two parameters characterizing the average efficiency of light use for dry matter production.

The amount of light intercepted by a particular species in a mixture is calculated from its leaf area index and the light profile within the canopy. In the most simple version, the light interception of a species is derived from the light intensity at half its plant height, which can be represented by the following equation, assuming that the leaf area is evenly distributed over plant height:

$$sh_i = \exp\left\{-\sum_{j=1}^{n}\left(k_j \, LAI_j \, \frac{h_{t,j} - 0.5 h_{t,i}}{h_{t,j}}\right)\right\} \quad \text{with } h_{t,j} - 0.5 h_{t,i} \geq 0 \quad (4.27)$$

in which sh_i represents the effective share of leaf area of species i. The growth rate of a species can then be calculated as:

$$G_{p,i} = \frac{k_i \, LAI_i \, sh_i}{\Sigma \, (k \, LAI \, sh)} \{1 - \exp(-\Sigma k \, LAI)\} \bullet E_{dm} \bullet I_0 \bullet (1 - \rho) \quad (4.28)$$

where
$G_{p,i}$ is the potential growth rate of species i (kg dry matter ha^{-1} d^{-1}),
E_{dm} the average light use efficiency (kg dry matter MJ^{-1}), and
I_0 the amount of incoming *PAR* (MJ m^{-2} ground d^{-1}).
Compared to the above described detailed CO_2 assimilation model, this relatively simple approach already gives a reasonable approximation.

Fig. 4.7 shows the difference between the methods for calculation of CO_2 assimilation of a mixed canopy with identical species only differing in height. The simple method gives a reasonable estimation of the relative light absorption of the species but the relative rates of CO_2 assimilation are overestimated when species strongly differ in height. This is the result of the non-linear CO_2 assimilation-light response curve of single leaves causing a different light use efficiency for leaves at the bottom of the canopy.

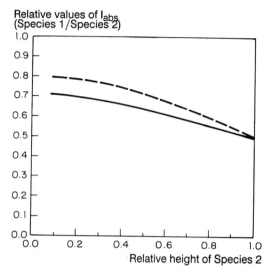

Fig. 4.7. Comparison of detailed and simple method for calculation of the share in light absorption. For two identical species with both $LAI = 2$, 100 and 300 J m^{-2} s^{-1} diffuse and direct radiation, respectively, the tallest species is plotted versus the relative height of the shortest species. Relative light absorption (I_{abs}) as calculated with the detailed model INTERCOM (solid line), and relative light absorption using the simple method (broken line).

Phenological development

The developmental stage of a plant defines its physiological age and is characterized by the formation of the various organs and their appearance. The most important phenological switch is the change from the vegetative to the reproductive stage. As many physiological and morphological processes change with the phenological stage of the plant, accurate quantification of phenological development is essential in any simulation model for plant growth. Temperature is the main driving force of developmental changes in potential production situations. However, in many so called photo-sensitive species and crop varieties, day length also determines induction of flowering.

Two approaches are followed in the model dependent upon the species and available data. The most simple approach is the widely used temperature sum approach, which assumes a linear relation between temperature and developmental rate, which is a realistic approach over a wide range of temperatures for many species (van Dobben, 1962; van Keulen and Seligman, 1987). The temperature sum or heat sum (*ts*, °C d) can be characterized as:

$$ts = \Sigma \, (T_{av} - T_b); \qquad T_{av} - T_b \geq 0 \qquad (4.29)$$

where

T_{av} is the average daily temperature (°C), and
T_b is the base temperature (°C), below which the phenological development stops.

The minimum value for $(T_{av} - T_b)$ is 0, as the phenological development is an irreversible process. For photo-sensitive varieties this approach can only be used for environments similar to the environment in which the experiments for determination of species characteristics have been conducted.

For many annual species, the developmental stage can be easily described using a dimensionless variable (*D*) which has the value 0 at emergence, 1 at flowering and 2 at maturity (van Keulen *et al.*, 1982). The developmental stage *D* is the integral of the developmental rate D_r (d^{-1}). This developmental rate is the inverse of the period required for completing a developmental unit (emergence - flowering; flowering - maturity). When a linear temperature response is used, the rate of development on a given day can be calculated according to:

$$D_r = (T_{av} - T_b) \, / \, \Sigma \, (T_{av} - T_b) \qquad (4.30)$$

where

T_{av} is the average daily temperature (°C), and
T_b is the base temperature (°C), below which the phenological development stops.

Often, the rate of development is non-linearly related to temperature, like in rice (Penning de Vries *et al.*, 1989). The rate of development (D_r) is then defined as a function of temperature or, when the temperature sum approach is used, the effective daily temperature, which is calculated from a non-linear relation, instead of $(T_{av} - T_b)$. In the model INTERCOM, a species specific relationship is defined.

Several approaches to quantify the photoperiodic effects on development have been developed (Weir *et al.*, 1984; Penning de Vries *et al.*, 1989; Miglietta, 1989, 1991a, b; Muchow and Carberry, 1989; Carberry, 1991). The model can be adapted to include day length effects on phenological development using one of these procedures.

Dry matter partitioning

For each of the competing species, the total daily produced dry matter is partitioned among the various groups of plant organs (leaves, stems, storage organs and roots) according to partitioning coefficients (pc, kg dry matter organ kg^{-1} dry matter crop) defined as a function of the phenological development stage (D) of the species:

$$pc_k = f(D) \qquad\qquad (4.31)$$

The dry matter is first distributed over shoot and root and then the shoot fraction is divided between stems, leaves and storage organs. This also holds for below ground storage organs like in sugar beet. The growth rate of plant organ group k ($G_{p,k}$) is thus obtained by multiplying the total potential growth rate (G_p, Eqn 4.26) by the fraction allocated to that organ group k (pc_k):

$$G_{p,k} = pc_k\, G_p \qquad\qquad (4.32)$$

Total dry weights of the plant organs are obtained by integrating their daily growth rates over time. This approach to dry matter partitioning can be improved by introducing effects of other factors that determine partitioning patterns like the water and nutrient status of the crop (van Keulen and Seligman, 1987). Some simulation models simulate a reserve pool of assimilates that enable the simulation of processes that determine source-sink relationships (de Wit *et al.*, 1978; Thornley, 1972; Ng and Loomis, 1984). This is not accounted for in the present version of the model INTERCOM, but can be included if it is needed to obtain accurate simulations of plant growth of certain species.

Redistribution of dry matter

The assimilates accumulated in storage organs partly originate from carbohydrates stored in stems and leaves before the grain filling starts. A very simple procedure is used in the model to redistribute dry matter from the stems to the storage organs: a fraction of the dying tissue is added to the total growth rate of the crop, the fraction being a function of development stage.

In the model INTERCOM, sink limitation is not accounted for. In grain crops the carbohydrate production in the grain filling period can be higher than the storage capacity of the grains, which is determined by the number of grains per m^2 and the maximum growth rate of the grains. This may result in the accumulation of assimilates in the leaves causing reduced rates of CO_2 assimilation through a feedback mechanism (Barnett and Pearce, 1983). This can be very important in rice when it is grown in extreme environments as both low and high temperatures before flowering can induce spikelet sterility which results in a low sink capacity (Yoshida, 1981). Several procedures to account for sink limitation have been developed and can be included in the model (van Keulen and Seligman, 1987; Spitters *et al.*, 1989a).

Morphological development: leaf area dynamics

The green leaf area of plants determines the amount of absorbed light and thus CO_2 assimilation. Especially in competition situations, an accurate simulation of leaf area development is required, because of the strong impact of *LAI* of a species on light capture in a mixed canopy (Fig. 4.6). In early versions of the model (Spitters and Aerts, 1983; Kropff *et al.*, 1984), leaf area development was assumed to be only determined by the amount of carbohydrates available for leaf growth. Leaf area was calculated from leaf dry matter using the Specific Leaf Area (*SLA*, m^2 leaf kg^{-1} leaf). This resulted in an extreme sensitivity of the model to the value of the parameter *SLA*, the physiological parameters related to CO_2 assimilation, and partitioning coefficients for the leaves, all of which appear to be highly variable in field conditions (Kropff, 1988a). Light intensity determines the rate of CO_2 assimilation and hence the supply of assimilates to the leaves, whereas temperature affects the rates of cell division and expansion.

Often, temperature is the overriding factor determining leaf area development during the early stage of plant growth (Horie *et al.*, 1979). Because the total *LAI* is small the leaves are not shading each other. All leaves absorb enough radiation to fulfil assimilate needs for leaf expansion. Then, in most species, leaf area increases more or less exponentially in time, with time expressed in terms of temperature sum:

$$LAI_{ts} = N L_{p,0} \exp(R_l ts) \qquad (4.33)$$

where
LAI_{ts} is the leaf area index (m^2 leaf m^{-2} ground) at a specific temperature sum (*ts* , °C d) after emergence,
N the number of plants per m^2,
$L_{p,0}$ the initial leaf area per plant at seedling emergence (m^2 plant^{-1}), and
R_l the relative leaf area growth rate (°C^{-1} d^{-1}).

The temperature sum *ts* is calculated according to Eqn 4.29 using a different base temperature if needed. The exponential phase ends when the portion of assimilates allocated to non-leaf tissue sharply increases, or when mutual shading becomes substantial. As a rule of thumb, one can use *LAI* (for total canopy) = 1 as the end of the exponential growth period, since leaves start to overlap by then. This can easily be checked by plotting ln*LAI* versus *ts* and analysing until when growth is linear.

The daily increase in leaf area per plant can be calculated from:

$$\frac{dL_p}{dt} = \frac{dts}{dt} \bullet R_1 \bullet L_{p,ts} \tag{4.34}$$

where
L_p is the leaf area per plant (m^2 leaf plant^{-1}),
ts the temperature sum (°C d), and
R_1 the relative growth rate of leaf area (°C^{-1} d^{-1})).
However, the model operates at a time step of one day, which is too long for this exponential process, causing severe underestimation of leaf area development when rectangular integration techniques are used (for discussion of time steps in integration we refer to Leffelaar and Ferrari, 1989). Therefore, a difference equation is used rather than a differential equation, because a difference equation gives the exact analytical integration of the equation over the day:

$$L_{p(t+dt)} - L_{p(t)} = L_{p,0} \exp\left(R_1(ts + \frac{dts}{dt})\right) - L_{p,0} \exp(R_1 ts) = L_{p,0} \exp\left(R_1 \frac{dts}{dt} - 1\right)$$

$$\tag{4.35}$$

where
dt is the time interval of integration, which is 1 day in the model,
ts the temperature sum since emergence (°C d), and
dts the effective temperature for that specific day.
In later stages of growth (*LAI*>1), leaf area increment is increasingly limited by assimilate supply as a result of an increasing amount of growing points, stem growth and mutual shading of leaves. Leaf area increment (*GLAI*, m^2 leaf m^{-2} ground d^{-1}) is then calculated by multiplying the simulated daily leaf weight increment (G_{lv}) of the species by the specific leaf area *SLA* (m^2 leaf kg^{-1} leaf) of new leaves:

$$GLAI = SLA \bullet G_{lv} \tag{4.36}$$

As the *SLA* of new leaves is difficult to derive from experimental data, the *SLA* of newly formed leaves can be estimated from the average measured *SLA* of all leaves.

To account for leaf senescence, a relative leaf death rate is defined, being a function of both the development stage of the species and the ambient temperature. Leaf death rate ($DLAI$, m^2 leaf m^{-2} ground ($^{\circ}$C d)$^{-1}$ or m^2 leaf m^{-2} ground d^{-1}) is then obtained by multiplying the green LAI by its relative death rate (R_s, ($^{\circ}$C d)$^{-1}$):

$$DLAI = LAI \cdot R_s \qquad (4.37)$$

The relative death rate is a species specific function of developmental stage. For the death rate a difference equation (see Eqn 4.35) is used to avoid bias as a result of rectangular integration, like in the procedure for early leaf area growth:

$$LAI_{t+dt} - LAI_t = LAI_t \exp\left(R_s \cdot \frac{dts}{dt} - 1\right) \qquad (4.38)$$

In the model INTERCOM, the relative death rate of the leaves is applied to the leaf weight. The death of leaf area is calculated from the loss of leaf weight using the SLA.

These simple approaches can be expanded in several ways. The specific leaf area of new leaves can be related to temperature and radiation (Acock *et al.*, 1978; Sheehy *et al.*, 1980; Jones and Hesketh, 1980). Several approaches to simulate leaf area development in a more mechanistic way have been developed (Stapper and Arkin, 1980; Jones and Hesketh, 1980; Weir *et al.*, 1984; Porter, 1984). In these approaches, leaf appearance is simulated as well as leaf expansion and the duration of leaf expansion to simulate leaf area development. Recently, several new approaches have been developed (Muchow and Carberry, 1989; Miglietta, 1989, 1991a, b), in which the final leaf number is simulated on the basis of photoperiod and temperature. For application of such approaches in competition situations, the feedback of dry matter growth rate on leaf area development has to be simulated explicitly. So far, we have had reasonable results by assuming leaf area development to be sink-limited before canopy closure and source-limited after canopy closure.

Morphological development: height growth

Plant height development can be described by a logistic function of the development stage or temperature sum (ts, $^{\circ}$C d) (Spitters, 1989):

$$h_{ts} = h_m / (1 + b \cdot \exp(-s \cdot ts)) \qquad (4.39)$$

where

h_{ts} is the actual height at time ts (m),
h_m the maximum height of the species (m),
b parameter of the logistic function (-), and
s parameter of the logistic function (($^{\circ}$C d)$^{-1}$).
The rate of height increment per day is given by the derivative of h_{ts}:

$$\frac{dh}{dt} = \left(\frac{dts}{dt}\right) \bullet s \bullet b \bullet h_m \exp(-s \bullet ts) / (1 + b \exp(-s \bullet ts))^2 \qquad (4.40)$$

where
dh/dt is the rate of height increment per day (m d^{-1}), and
dts/dt is the temperature sum for that day.
Height is obtained by integration of the height growth rate.

In competitive situations, however, height growth can be strongly reduced by shortage of assimilates as a result of shading. In crop-weed competition studies, this is especially the case for late emerging weeds. In the model, an empirical function was introduced that limits the Specific Stem Length (SSL, m height kg^{-1} stem) to a maximum value. This maximum value depends on plant height, i.e. a short plant can be thinner than a tall plant. In this way, shortage of assimilates as a result of competition has feedback on height growth. The function relates the maximum SSL to plant height (h):

$$SSL_{max} = \exp(-a\ h\ + c) \qquad (4.41)$$

where
SSL_{max} is the maximum SSL possible for a species (m kg^{-1}), and
a and c are constants.

If the amount of assimilates available for stem growth on a specific day is not enough to meet the requirements for the potential height growth and the maximum SSL has been reached (thinnest stem possible) (Eqn 4.40), height growth on that day is set to 0. This function defines the phenotypic plasticity of a species, with respect to height growth.

Summary and discussion

Because of the complex nature of competition for light, process based models can help to understand the system and to identify characteristics that determine competitiveness. In this chapter, the detailed approach that is used in the model INTERCOM has been described.

The approach to quantify competition for light in the model INTERCOM accounts for vertical heterogeneity of processes in the canopy, but assumes a horizontally homogeneous distribution of leaf area. For most weedy crop canopies this assumption will be valid. Horizontal heterogeneity can be accounted for in the model by distinguishing smaller patches in the field and running the simulation model for the different patches separately, assuming homogeneity within the patch. For row-structured canopy situations, like in intercropping systems, the detailed light absorption model for row crops developed by Goudriaan (1977) and Gijzen and Goudriaan (1989) could be adapted for rows consisting of different species.

Chapter Five

Mechanisms of Competition for Water

M.J. Kropff

Introduction and overview of processes involved

Water is one of the most important resources needed for crop growth. In contrast to the resource light, water supply is often manageable by irrigation systems. Many agricultural systems have been designed to optimize water supply of the crop. Especially in lowland rice ecosystems of which 63% of the area can be classified as irrigated (IRRI, 1989), minimization of water use or optimization of the water use efficiency is a major concern for farmers and scientists. Rainfed environments are generally highly variable. The factors that determine crop growth in these environments are the seasonal water supply, soil water holding capacity, root development and the water use efficiency. Plants require water for transpiration via stomata. If there is a shortage in water supply, transpiration requirements cannot be met by water uptake by the roots, causing the closure of the stomata. Because CO_2 is also transported through the stomata, growth is then reduced as well.

Plants growing in a mixture in a water limited production situation compete for water and generally also for light. The importance of water competition is determined by the length, severity and timing of the drought period(s). Competition mechanisms for water differ principally from competition for light. Competition for light is a process of direct competition for resource capture, with an instantaneous nature: if the resource is not captured, it is lost because the resource is not stored in the system. The amount of absorbed radiation generally limits photosynthesis of closed leaf canopies in potential production situations. Light absorption by neighbouring plants reduces the rate of CO_2 assimilation and growth of the plant.

Water can be stored in the system in contrast to the resource light. However, it is only stored in relevant quantities in the soil compartment of the system. So, when the soil supply cannot meet the demand, the crop experiences effects immediately. When plants compete for water, two processes have to be distinguished. The first process is direct competition for water. In a dry period when water is limiting growth, competition for resource capture is an instantaneous process. The plant that has better access to soil moisture (deeper rooting system, higher root density) has an instantaneous benefit. The second process can be characterized as an indirect effect. In periods where soil moisture content is high enough for potential growth, all species in a mixture can meet their water requirements for transpiration during that period. This is often the case during the beginning of the growing season. These requirements differ between species because they are determined by the amount of absorbed radiation, temperature, vapour pressure deficit and species characteristics. However, the amount of available soil moisture will be reduced during the growing season, when rainfall and other processes that increase soil moisture content cannot meet the water losses by evapotranspiration and drainage. Therefore, plant transpiration in a period when water is not limiting growth affects the growing situation later in the season. Especially when the life cycle of species in a mixture differs, it may happen that an early maturing species does not suffer from water stress itself, but increases the water stress effects for the later maturing species by enhancing total water loss earlier in the season. For example, in one of our experiments with maize and *Echinochloa crus-galli* L., severe drought occurred during stem elongation of the maize. The competing *E. crus-galli* plants almost completed their life cycle by then and were hardly affected by the drought. Water competition effects on maize, however, were very strong, as the mixture transpired more water than the monoculture maize that had an open canopy for a long period, but where a crust on the soil prevented evaporation. This example will be discussed in detail in Chapter 8. Species differ in their water use efficiency with respect to dry matter production, which affects competitive relationships (Radosevich and Holt, 1984). It is well known that C_4 species require less water per unit dry matter produced when compared to C_3 species. A more efficient species may have an advantage in a drought situation, but often the dynamics and timing of drought periods determine if that holds. Radosevich and Holt (1984) pointed out that the ability of a species to survive stress and the ability to outcompete neighbours or control resource availability are not necessarily the same phenomenon.

Because both light and water competition (direct and indirect) are important in water limited production situations, competition effects have to be studied in an integrated way. That can be done by simulation modelling. Experimental analyses are extremely complex because of the complex dynamics of the system. By incorporating modules that simulate processes related to the water balance in the canopy and the soil in the competition model INTERCOM, the complex interactions can be studied. The major processes that are involved are transpiration, evaporation from the soil, effects of water stress on transpiration and growth, and other processes that determine the soil water balance like percolation and capillary rise.

In this chapter, we will discuss the soil water balance, the processes that determine changes in the water balance of the soil, and the effect of water shortage on plant growth in a mixture as simulated by the model INTERCOM. Ways to increase comprehensiveness of the model INTERCOM are indicated.

Soil water balance

For dryland conditions, a simple moisture balance for a free draining soil profile is included in the model INTERCOM (Subroutine WBAL, Appendix 2) (Fig. 5.1A). Two horizontal soil layers are distinguished: a 2 cm top soil layer and the rooted soil layer, bounded by the top soil layer and the rooted depth of the soil. The 2 cm top soil layer was included to enable realistic simulation of soil evaporation in periods when the top soil dries out and evaporation is reduced because of crust formation. The maximum rooted depth is used to define the lower boundary of the second layer throughout the growing season. The daily change in soil moisture content in the top soil layer ($d\theta_1$, kg H_2O m^{-2} d^{-1}) is calculated according to:

$$d\theta_1 = R - E1 - P1 + IR \qquad (5.1)$$

where
R is the rainfall (kg H_2O m^{-2} d^{-1}),
$E1$ the soil evaporation of the top layer (kg H_2O m^{-2} d^{-1}),
$P1$ the percolation to the second layer (kg H_2O m^{-2} d^{-1}), and
IR the irrigation (kg H_2O m^{-2} d^{-1}).
Daily data on rainfall (in mm) are input for the model. Percolation is calculated as the amount of water in excess of field capacity.

The change in soil moisture content of the rooted zone ($d\theta_2$, kg H_2O m^{-2} d^{-1}) is calculated as:

$$d\theta_2 = P1 + CR - T_a - E2 - P2 \qquad (5.2)$$

where
$P1$ is the percolation from the top soil layer to the second layer (kg H_2O m^{-2} d^{-1}),

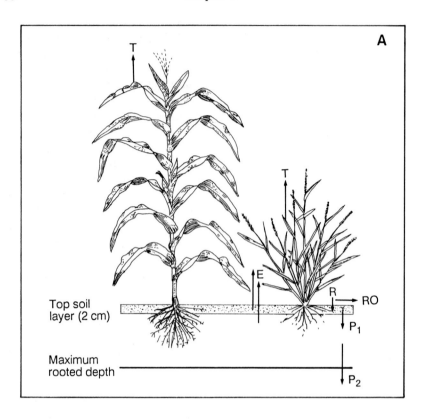

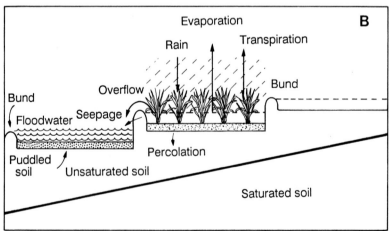

Fig. 5.1. (A) Components of the soil water balance for a free-draining soil (T, Transpiration; E, Evaporation; R, Rainfall; RO, Run-off; P, Percolation). (B) Components of the water balance of rainfed lowland rice fields. Source: Angus, 1991.

CR   the capillary rise ((kg H_2O m^{-2} d^{-1}),
T_a the actual transpiration by the vegetation (kg H_2O m^{-2} d^{-1}),
E2   the soil evaporation (kg H_2O m^{-2} d^{-1}), and
P2   deep percolation (kg H_2O m^{-2} d^{-1}).

Daily data on rainfall (in mm) are input to the model. Percolation is calculated as the amount of water in the layer in excess of field capacity, which drains below the rooted zone with a delay of a few days. This delay is introduced in the model by a time coefficient for percolation, which depends on the soil type. Capillary rise is input to the model as a function, but a subroutine that simulates capillary rise could be added to the model. Total evaporation is distributed over the two layers, based on an empirical distribution factor (0.26 from the total daily evaporation is assumed to come from the top layer). This factor was derived from simulation studies with the model of van Keulen (1975), that calculates distribution of evaporation over a large number of soil layers according to an exponential function.

In earlier versions of the model, a multi layer model with one top soil layer and a large number of layers (generally 10) with equal thickness was used (developed by van Keulen (1975)). The SAHEL module for the soil water balance presented by Penning de Vries *et al.* (1989) is also based on this principle and can easily be coupled to the model INTERCOM. The simple approach presented here accounts for the indirect competition effects and only partly for the direct competition effects. The effects of differences between rooted depth of species, and differences in the root length density distribution with depth are not accounted for. These effects can easily be included in the model. However, detailed data and insight will be required for such approaches (e.g. the relationship between root length density and water uptake capacity cannot be quantified yet).

For lowland conditions in rice, or in poorly drained soils, where the groundwater table is between 0 - 1 m below soil surface during part of or the entire growing season, a completely different approach has to be taken (Fig. 5.1B). Capillary rise has to be simulated explicitly. The module SAWAH, developed Penning de Vries *et al.* (1989) can be coupled to the competition model to simulate soil moisture contents for paddy rice fields. The SAWAH module calculates water flow between soil layers on the basis of hydraulic properties of the different soil layers.

Potential evapotranspiration

The potential rate of soil evaporation and crop transpiration is calculated from the reference evapotranspiration of a short grass cover

(ET_r), which is calculated from the daily weather data (vapour pressure, temperature, wind speed) using a Penman type equation (Penman, 1948) (Subroutine PENMAN, Appendix 2). In that approach, the effect of radiation that provides the energy for water evaporation is considered in combination with the effect of turbulence in the air to remove the water vapour. This procedure enables the calculation of potential evapotranspiration from data obtained from standard meteorological stations.

Table 5.1. Indicative values for the empirical constants a_A and b_A in the Ångström formula, in relation to latitude used by the Food and Agriculture Organization, FAO. Source: Frère and Popov, 1979.

	a_A	b_A
Cold and temperate zones	0.18	0.55
Dry tropical zones	0.25	0.45
Humid tropical zones	0.29	0.42

Net radiation

Total global daily incoming is measured only at a limited number of meteorological stations. For other stations, total global radiation (S_g, J m^{-2} d^{-1}) can be estimated from measured sunshine duration using an empirical function (the Ångström formula):

$$S_g = S_0 \bullet (a_A + b_A \bullet (n_s/N_s))\qquad(5.3)$$

where
S_0 is the theoretical amount of global radiation without an atmosphere,
a_A an empirical constant (see Table 5.1),
b_A an empirical constant (see Table 5.1), and
n_s/N_s the ratio between the amount of bright sunshine hours (n_s) and the maximum amount of sunshine hours (N_s).

The earth's surface and the atmosphere have temperatures above the absolute zero temperature (K) and emit thermal or long-wave radiation. As the temperature of the earth's surface is higher than the temperature of the atmosphere, there is a net flux of outgoing long-wave radiation from the earth's surface.

The net outgoing radiation (R_b, J m^{-2} d^{-1}) can be estimated from an equation similar to the one proposed by Brunt (1932) (Penman, 1956):

$$R_b = \sigma_{sb}\left(T_{av} + 273\right)^4 \bullet \left(0.56 - 0.079\sqrt{e_a}\right) \bullet \left(0.1 + 0.9\,\frac{n_s}{N_s}\right) \qquad (5.4)$$

where

σ_{sb} is the Stefan-Boltzmann constant (J m^{-2} d^{-1} K^{-4}),
T_{av} the average daily temperature during daytime ($^{\circ}$C),
e_a the actual vapour pressure (mbar), and
n_s/N_s the ratio of actual and maximum hours of sunshine, which is calculated from the atmospheric transmission (Eqn 5.3) using the Ångström formula

$$\frac{n_s}{N_s} = \frac{S_g/S_0 - a_A}{b_A} \qquad (5.5)$$

The net radiation (R_n, J m^{-2} d^{-1}) is calculated according to:

$$R_n = (1 - \rho_g) \bullet S_g - R_b \qquad (5.6)$$

where

ρ_g is the reflection coefficient of the surface for global radiation, or the albedo (-),
S_g the global incoming radiation (J m^{-2} d^{-1}), and
R_b the net outgoing long-wave radiation emitted by the earth (J m^{-2} d^{-1}).

Indicative albedo values are: for a water surface 0.05, for a soil surface 0.15 and for a crop surface a value of 0.25 is used (Penman, 1956).

Reference evapotranspiration

The combination equation of Penman, which is used to calculate the reference evapotranspiration, can be derived from the radiation balance equation for an extensive area of open water, wet soil or crop, assuming that the net energy transfer between the sky and the soil, crop or water surface on a daily basis equals 0:

$$R_n = H + \lambda E \qquad (5.7)$$

where

R_n is the net radiation (J m^{-2} d^{-1}),
H represents the sensible heat flux between surface and air (J

m^{-2} d^{-1}), and
λE is the potential evapo(transpi)ration or latent heat flux from the surface to the air (J m^{-2} d^{-1}).

The sensible heat loss of a surface (H) is proportional to the temperature difference and a heat transfer coefficient, which depends on wind speed. In analogy, the latent heat loss, through loss of water vapour at the surface, is proportional to the difference in vapour pressure between surface and the surrounding air. As potential evapotranspiration is considered, the air at the surface is water vapour saturated. The saturated vapour pressure of air (e_s) can be approximated with an empirical function (Goudriaan, 1977):

$$e_s = 6.11 \bullet \exp(17.4\, T / (T + 239)) \tag{5.8}$$

where T is the temperature of the air (°C). Because the temperature at the surface is unknown, Penman (1948) introduced the method of linearizing the saturated vapour pressure curve (characterized with the slope s, mbar °C^{-1}). The set of equations for the energy balance can then be solved, resulting in the combination equation for the calculation of evapotranspiration. The combination equation to calculate potential evapotranspiration (Penman, 1948) can be written as the sum of two 'forces' driving the evapo(transpi)ration: a radiation term λE_s and an aerodynamic term λE_d:

$$\lambda E = \lambda E_s + \lambda E_d = \frac{sR_n}{s + \gamma} + \frac{\lambda \gamma h_u (e_s - e_a)}{s + \gamma} \tag{5.9}$$

where
λE_s is the radiation driven component of λE (J m^{-2} d^{-1}),
λE_d the air driven component of λE (J m^{-2} d^{-1}),
R_n the net radiation (J m^{-2} d^{-1}),
s represents the slope of saturated vapour pressure curve at air temperature (mbar °C^{-1}),
λ the latent heat of vaporization of water (J kg^{-1}),
γ the psychrometer constant (mbar °C^{-1}),
h_u the wind function (kg H$_2$O m^{-2} d^{-1} mbar^{-1}),
e_s the saturated vapour pressure at air temperature (mbar), and
e_a the actual vapour pressure (mbar).

The wind function h_u estimates the conductance for transfer of latent and sensible heat from the surface to the reference height (2 m) and is a function of wind speed at 2 m height (u_2, m s^{-1}). In the Penman method most often empirical wind functions are used that are implicitly parameterized for effects of roughness of the surface and atmospheric stability. Wind functions that are physically more sound are also available (Goudriaan, 1977). The wind function that is mostly used for open water and soil surfaces is (Penman, 1956):

$$h_u = 0.263 \bullet (0.5 + 0.54u_2) \tag{5.10}$$

with h_u in kg H_2O m^{-2} d^{-1} mbar^{-1}, and u_2 in m s^{-1} at 2 m height. This formula has been modified to conform to the units which explains the difference with the expressions for open water found normally in the literature. The wind function that is mostly used for short, closed grass crops is:

$$h_u = 0.263 \bullet (1 + 0.54u_2) \tag{5.11}$$

Soil evaporation

The actual rate of soil evaporation is a function of the potential evaporation rate, the hydraulic conductivity of the soil and the fraction of radiation transmitted through the canopy and absorbed by the soil (this is simulated in the Subroutine DEVAP, Appendix 2). The radiation intercepted by the vegetation and the radiation transmitted to the soil are derived from the exponential radiation profile in the canopy (Eqn 4.5). That gives for the potential soil evaporation, assuming a wet surface (E_p, kg H_2O m^{-2} d^{-1}):

$$E_p = E_{p,r} \bullet \exp(-\Sigma \, (0.7 \, k_j \, LAI_j)) \tag{5.12}$$

where
$E_{p,r}$ is the soil evaporation without a crop (calculated with Eqn 5.9 using an albedo of 0.15 and the wind function given in Eqn 5.10) (kg H_2O m^{-2} d^{-1}),
k the extinction coefficient for *PAR* (-), and
0.7   the ratio between the extinction coefficient for total global radiation and that for *PAR*.

This factor 0.7 results from the high scattering coefficient for near infra red radiation (0.2 for *PAR*, 0.8 for *NIR*). *LAIs* are weighted by the species respective extinction coefficients and summed over the j = 1,...,n species constituting the mixture.

    Water shortage in the top layer reduces the rate of evaporation through a reduced hydraulic conductivity. The actual value of the soil evaporation (E_a) is obtained by multiplying the potential value (E_p) by a reduction factor (E_{rd}) which is a function of the moisture content of the top 2 cm of the soil (van Keulen, 1975; van Keulen and Seligman, 1987) (Fig. 5.2). The actual evaporation is partly taken

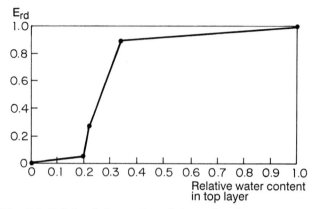

Fig. 5.2. Relation between the relative soil moisture content in the top soil compartment, and the reduction for soil surface evaporation, E_{rd}. Source: van Keulen and Seligman, 1987.

from the top soil layer (26% by the top layer and 74% by the rooted layer), based on a simplification of the model developed by van Keulen (1975).

Transpiration of a mixed canopy

The rates of potential and actual transpiration are calculated in the Subroutine TOTRAN (Appendix 2).

Potential transpiration

The potential rate of transpiration is derived from the reference evapotranspiration of a short grass cover (ET_r), calculated from the weather data using the Penman (1948) equation (Eqn 5.9) using the wind function presented in Eqn 5.11 and an albedo of 25%.

Rates of potential transpiration are approximately proportional to the amount of total solar radiation intercepted by the competing species. The fraction of radiation intercepted by the competing plants ($f_{a,i}$) is derived from the detailed mixed canopy light absorption subroutines as used in the model INTERCOM (Subroutines TOTASS, RADIAT, ASTRO, ASSIMC). The fraction of *PAR* absorbed by the different species, instead of total global radiation (50% *PAR* + 50% *NIR*) is used in the model version of INTERCOM presented here for reasons of simplicity. However, a subroutine that calculates absorp-

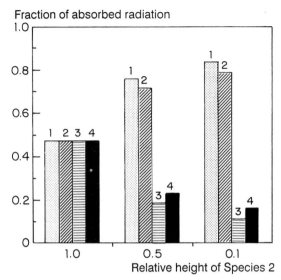

Fig. 5.3. Fraction of radiation absorbed by two competing species (*LAI* = 3, at different relative heights) as simulated with two models for light competition (ASSIMC): one simulates absorption of *PAR* and one simulates absorption of global radiation by distinguishing *PAR* and near infra red radiation (*NIR*). Fraction absorbed *PAR* for Species 1 (1) and Species 2 (3); fraction absorbed global radiation by Species 1 (2) and Species 2 (4).

tion of *PAR* and *NIR*, separately, is available on request. The scattering coefficient for *NIR* is much higher than for *PAR* (0.8 for *NIR*; 0.2 for *PAR*) resulting in a much lower (0.7) extinction coefficient for *NIR*. The resulting strong difference in radiation profiles for *PAR* and *NIR*, however, does not cause strong differences between the fraction of *PAR* and *PAR* + *NIR* absorbed, even when the species strongly differ in plant height (Fig. 5.3). Thus, the potential rate of transpiration for species i ($T_{p,i}$, mm d^{-1}) can be approximated by:

$$T_{p,i} = ET_r \cdot f_{a,i} \cdot c_i \tag{5.13}$$

where

ET_r is the reference evapotranspiration for a short grass cover (mm d^{-1}),

$f_{a,i}$ the fraction of total incoming radiation (*PAR*) absorbed by species i (calculated in the Subroutine TOTASS, Appendix 2), and

c_i is an empirically determined proportionality factor the 'crop factor' that compares potential evapotranspiration of a short grass cover with a crop canopy (closed).

This crop factor is about 0.9 for C_3 species (Feddes, 1987). For C_4 species a value of 0.7 is set for temperate climates, where C_4 species hardly grow faster than C_3 crops, but have a much higher water use efficiency. In tropical climates, C_4 crops can grow almost twice as fast as C_3 crops resulting in a value of 0.9 for these species as well.

Actual transpiration

Water shortage reduces the rate of transpiration. The ratio between actual (T_a) and potential transpiration (T_p) decreases linearly with soil moisture availability when the actual soil moisture content (θ_a) falls below a certain critical level (θ_{cr}) (Doorenbos and Kassam, 1979) (Fig. 5.4):

$$T_a/T_p = (\theta_a - \theta_{wp}) / (\theta_{cr} - \theta_{wp}); \quad \theta_{wp} \le \theta_a \le \theta_{cr}$$

$$T_a/T_p = 1; \quad \theta_a > \theta_{cr} \tag{5.14}$$

where the critical soil moisture content is defined as

$$\theta_{cr} = \theta_{wp} + (1-p)(\theta_{fc} - \theta_{wp}) \tag{5.15}$$

in which θ is the soil moisture content (kg H_2O m^{-2} ground or mm), subscripts denoting the critical value (cr) and the values at wilting point (wp) and field capacity (fc), respectively. The soil moisture depletion factor (p) depends on plant species and evaporative demand (Fig. 5.4). For C_3 species it typically varies between 0.6 and 0.4 at an evaporative demand of 1 and 5 mm d^{-1}, respectively.

Effects of water shortage on growth processes

The reduction in growth rate is more or less proportional to the reduction in transpiration rate (van Keulen, 1975). Thus, the actual growth rate G_a, limited by soil moisture, is obtained by multiplying its potential value $(G_p$; Eqn 4.26 or 4.28) by the factor T_a/T_p (see Subroutine PLANTC variable TRANRF, Appendix 2):

$$G_a = T_a/T_p \bullet G_p \tag{5.16}$$

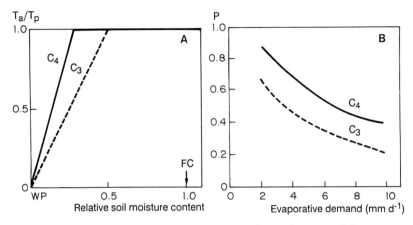

Fig. 5.4. (A) The relationship between the ratio of actual to potential transpiration (T_a/T_p) and the relative soil moisture content (WP, wilting point; FC, field capacity) and (B) the relationship between the soil moisture content (P) and the evaporative demand (mm d^{-1}) for a C$_3$ and a C$_4$ plant. Redrawn after Doorenbos and Kassam (1979).

The multiplication factor T_a/T_p is also applied to the rate of plant height increment and early leaf area growth.

Drought also affects rates of leaf expansion and senescence, the dry matter partitioning between above-ground and below-ground parts, and the photosynthetic capacity of the leaves. However, detailed physiological knowledge for the species studied so far is not (yet) available. Therefore, these effects are not included in the model yet. This will be discussed in Chapter 8, where the model is evaluated with data from field experiments.

Discussion

The model INTERCOM presented here, mainly deals with indirect competition for water and not with the direct interaction assuming that the rooted depth of the species do not differ. However, sometimes plants have a significantly different rooting depth. In that situation a multi compartment soil moisture model has to be used, because the soil compartments from which the plants can extract water is the same for the species. Simulation of direct competition for water in a soil compartment requires more quantitative insight in these processes. In the current model, the transpiration requirements of the species are calculated based on their share in absorbed radiation and actual transpiration is simulated by using a reduction function which is a fraction of soil moisture content. The root length density of the species in a specific layer could be a determinant for the share of water required by that species as well. However, it was concluded by van Keulen (1975) that the simple assumption, that plants can extract water from a soil layer if it has roots in the layer, works well in drought situations.

Transpiration in the model INTERCOM is calculated using the Penman (1948) equation. A modified version of this equation for leaf transpiration was introduced by Monteith (1965), in which stomatal conductance for water vapour is included. However, the influence of that correction is relatively small, thus the original Penman equation gives a fair approximation of transpiration losses by crops (van Keulen and Wolf, 1986).

Chapter Six

Mechanisms of Competition for Nitrogen

M.J. Kropff

Introduction

In many production situations, the supply of macro-nutrients (N,P,K) limits the growth of the crop or vegetation for at least part of the growing season. A mechanistic understanding of the soil and plant processes related to availability, uptake and use of these nutrients is difficult to obtain. Mechanistic simulation models at the level of detail of the model INTERCOM have been developed for nitrogen effects on monoculture crops, but their predictive capability is still limited (cf. van Keulen and Seligman, 1987). Fourteen detailed models for N turnover in the soil-crop system were compared recently at a workshop (Groot *et al.*, 1991). It was concluded that modelling the soil N balance on a day to day basis is still problematic (de Willigen, 1991). For example, none of the models could account for losses of mineral N occurring shortly after fertilizer application. Relatively simple approaches are, however, available (cf. Janssen *et al.*, 1990) and may be used to obtain more insight in processes related to competition for nutrients between plants. These models operate with time steps of a cropping season and have to be adapted to account for competition effects. Several models for nutrient competition have been developed to analyse long-term changes in (semi)natural plant communities (Tilman, 1988; Berendse *et al.*, 1987; Berendse *et al.*, 1989). These models operate at time steps of one year. The level of detail at which these models operate is, therefore, not suitable to understand the 'within' season dynamics in crop-weed systems.

In the model INTERCOM, competition for nutrients is not incorporated, because the model was only evaluated for production situations where nutrients were available in ample supply. However, in one of our simulation studies, it was hypothesized that nitrogen uptake by the weeds early in the growing season caused deviations of simulated versus observed yield losses by weeds in tomatoes (Weaver *et al.*, 1992).

In this chapter, a simple approach for simulation of indirect and direct competition for nitrogen is introduced that can be included in the model INTERCOM. The approach is based on the model of van Keulen (1982) for the impact of N fertilization on crop growth and production in monocultures, and was described by Spitters (1989). Similar approaches can be followed for the other nutrients.

Available soil nitrogen

In the simple approach presented here, the rooted zone of the soil profile is regarded as a single compartment in which all mineral nitrogen (nitrate, ammonium) is potentially available for uptake by the plants. At the beginning of the growing season an initial amount of mineral N will be available in the soil (N_0, kg N ha^{-1}). During the growing season, small amounts of mineral nitrogen become available as a result of mineralization of soil organic matter. In a competition situation, the species with the greatest nitrogen uptake capacity (i.e. largest root system) will benefit most from this release. Modelling all processes that determine soil N dynamics is complex and requires detailed information on soil characteristics (for reviews of models see Frissel and van Veen, 1981; de Willigen and Neeteson, 1985). In the approach presented here, net mineralization, which is the difference between mineralization and immobilization, is accounted for in a very simple way according to Greenwood *et al.* (1984). The daily rate of change of available soil N (N_s, kg N ha^{-1}) can be quantified by:

$$dN_s = M + F - U/r \qquad (6.1)$$

where
M is the net mineralization rate (kg N ha^{-1} d^{-1}),
F the fertilizer rate (kg N ha^{-1} d^{-1}),
U the N uptake by the vegetation (kg N ha^{-1} d^{-1}), and
r the recovery of N fertilizer (ratio of change in N uptake and change in N fertilizer rate).

Mineralization mainly depends on the amount of fresh organic material that can be easily mineralized. If the rate of mineralization is unknown, a simpler approach can be used, in which it is assumed that all mineral N, taken up by a crop grown without fertilizer, is available from the beginning of the growing season onwards. From standard nitrogen trials the recovery of fertilizer nitrogen can be calculated as the amount of extra N taken up divided by the rate of N fertilizer. The rate of change in N_s then equals the rate of N uptake.

Total available N in the soil is the measured N uptake in zero-N trials plus the amount of N fertilizer multiplied by the recovery. This approach does not take into account the dynamics of the soil N supply.

Nitrogen uptake

The potential rate of nitrogen uptake of the vegetation (D_N, demand in kg N ha^{-1} d^{-1}) equals the maximum amount of nitrogen in the crop, which can be determined from the maximum nitrogen concentration in the organs of the crop and the dry matter (van Keulen and Seligman, 1987), minus the actual amount:

$$D_N = (N_{c,m} - N_{c,a}) / T_c = (W \bullet NC_m - N_{c,a}) / T_c \qquad (6.2)$$

where
$N_{c,a}$ is the actual amount of N in the crop (kg N ha^{-1}),
$N_{c,m}$ the maximum amount of N in the crop (kg N ha^{-1}),
T_c the time coefficient (d),
NC_m the N concentration of the crop (kg N kg^{-1} dry matter), and
W the biomass (kg dry matter ha^{-1}).
The time coefficient T_c accounts for a delay in uptake, which is of the order of 2 days (Seligman and van Keulen, 1981).

The actual uptake of N by the crop (dN_c) is equal to the minimum of the demand (D_N) or the maximum supply by the soil (N_s).

$$dN_c = \text{Min}\,(D_N, N_s) \qquad (6.3)$$

Growth reduction

When the N content of the vegetation decreases below a certain level (NC_{cr}, the critical nitrogen concentration), the growth rate is reduced. This growth rate is assumed to be linearly related to the N concentration of the crop:

$$\frac{G_a}{G_p} = \frac{NC_a - NC_{mn}}{NC_{cr} - NC_{mn}} \qquad 0 \leq \frac{G_a}{G_p} \leq 1 \qquad (6.4)$$

where
G_a is the actual growth rate (kg dry matter ha^{-1} d^{-1}),
G_p the potential growth rate (kg dry matter ha^{-1} d^{-1}),
NC_a the actual N concentration of the crop (kg N kg^{-1} dry matter),

NC_{cr} the critical nitrogen concentration of the crop (kg N kg^{-1} dry matter), and

NC_{mn} the minimum nitrogen concentration of the crop (kg N kg^{-1} dry matter).

The actual N concentration (NC_a) is obtained by dividing the actual amount of N in the crop (N_c) by the biomass present (W_t).

The transpiration rate has to be reduced by the same factor (G_a/G_p) as the growth rate when nutrients are in short supply, because of stomatal control mechanisms (Goudriaan and van Keulen, 1979; Wong *et al.*, 1985).

Maximum, critical and minimum nutrient contents

The maximum and minimum nutrient contents differ strongly between plant organs, decrease with development stage, and vary among plant species. For nitrogen, typical values for these contents for field-grown annual C_3 grasses have been derived by van Keulen (1982, p. 240).

For annual C_3 grasses, Spitters (1989) derived values for maximum, minimum and critical N contents of the total above-ground biomass. The maximum N content of the above-ground parts decreases from 0.050 (kg N kg^{-1} dry matter) at emergence to 0.025 at flowering, and to 0.018 at maturity. The critical content equals 65% of the maximum content; and the minimum content is about 0.008 kg N kg^{-1} dry matter.

The maximum N content of C_4 grasses is probably close to those of the C_3 grasses. However, critical and minimum N contents of C_4 species are 50% of the values for C_3 species (Penning de Vries and van Keulen, 1982; Brown, 1985), because the C_4 pathway enables CO_2 assimilation to continue at lower levels of Rubisco (Ku *et al.*, 1979). Maximum N contents of legumes are 30% higher than those of grasses and for non-leguminous dicotyledons values are about 10% higher than those of grasses. Maximum N contents in seeds may differ strongly between species. The importance of the difference in minimum N contents between C_3 and C_4 species is indicated in Fig. 6.1, where the effect of this difference for two species that are identical for all other traits is simulated using a very simple competition model (Spitters, 1989).

Phosphorus (P) contents are coupled to nitrogen contents. P/N ratios vary within the range of 0.04 to 0.15 and have a value of 0.10 with ample nutrient supply (Penning de Vries and van Keulen, 1982). At very low levels of N uptake, absorption of P is restricted so that the P/N ratio does not exceed 0.15. At very low P availability, uptake of N

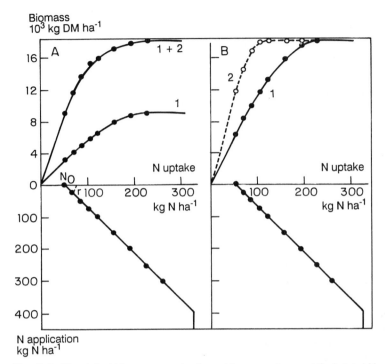

Fig. 6.1. Simulated biomass production of two species and their total N uptake in 1:1 mixture (A) and in a monoculture (B). The lower the soil N availability the greater the relative advantage of the C_4 type (Species 2) over the C_3 type (Species 1). At low N availability, the C_4 type utilizes the available N almost twice as efficiently as the C_3 type because it is able to dilute the N concentration in its tissues to concentrations that are half as low. The fertilizer recovery (r) equals 0.7 and the N uptake from unfertilized soil (N_0) is 50 kg N ha^{-1}. Source: Spitters, 1989.

is reduced so that the P/N ratio does not decrease below a value of 0.04. Thus, by combining these features with the N model discussed above, a simple model for P can be defined.

Potassium (K) contents are of roughly the same order of magnitude as N contents, but vary greatly among plant species and soil characteristics.

Competition for nutrients

In production situations where mobile soil elements like water and nitrates are not limiting growth, root length density has hardly any effect on the total uptake of these elements by the crop (Seligman and van Keulen, 1981; van Noordwijk, 1983). Therefore, the total uptake by a mixed vegetation can be calculated in a similar way to the procedure described for the monoculture situation.

However, when supply from the soil cannot cover demand, uptake by a species in a mixed vegetation will be related to its share in the total effective root length. Below-ground competition for soil elements is modelled in analogy with competition for light. The fraction of nutrient ions that is taken up by a species is related to its share in the total root system:

$$dN_{c,i} = (l_i/\Sigma l)\ \Sigma dN_c \qquad dN_{c,i} \le D_{N,i} \qquad (6.5)$$

where
$dN_{c,i}$ is the N uptake by species i (kg N ha^{-1} d^{-1}),
ΣdN_c the N uptake summed over all species of which the vegetation is composed (kg N ha^{-1} d^{-1}),
$D_{N,i}$ the N demand of species i (kg N ha^{-1} d^{-1}), and
l_i the effective root length of species i (m ha^{-1}).
It is important to note that the relative, rather than the absolute, effective root length of a species determines its competitive ability. This approach could also be used for the water competition part.

A species with an extensive root system, relative to its demand, is able to meet its demand up to a lower soil nitrogen supply. At limited soil supplies, the soil nitrogen which is not used by such a species is distributed over the other species.

The simple model was used to study competition between grass species in permanent grassland in long-term fertility trials in the Netherlands. This preliminary study showed that the simple approach, based on the assumption that all N is available at the beginning of the growing season, simulated N uptake and yield reasonably. However, a thorough evaluation of the approach is needed for replicated trials in different situations to prove its validity (Verberne, Elberse, Kropff and Lantinga, Department of Theoretical Production Ecology, Wageningen Agricultural University, unpublished data).

Chapter Seven

Eco-Physiological Characterization of the Species

M.J. Kropff and L.A.P. Lotz

Introduction

In the previous chapters, mechanistic approaches to understanding interplant competition were discussed using the framework of the model INTERCOM. To use these approaches to improve understanding of competitive phenomena based on process knowledge, it is important to determine genetic variations in eco-physiological and morphological traits. It is this variation that causes differences in competitiveness of species and cultivars. Because of the mechanistic nature of the approach used in the model INTERCOM, eco-physiological and morphological characteristics have to be quantified. The model can then be used to integrate these characteristics and to simulate competitive processes. This chapter will specifically deal with the procedures that have to be followed to estimate the parameter values. Detailed examples will be given for several crop and weed species: rice, maize, sugar beet, and the weeds *Chenopodium album* L. and *Echinochloa crus-galli* L.

Light profile in the canopy

The fraction of the downward radiation flux that is reflected by a green leaf canopy with a spherical leaf angle distribution (the reflection coefficient ρ) is calculated from the scattering coefficient (σ, Eqn 4.4). Goudriaan (1977) estimated the scattering coefficient of single leaves for visible radiation to be 0.2. This value is used as a standard for the calculation of CO_2 assimilation of crop canopies. The light profile in the canopy is determined by the extinction coefficient for visible radiation (k) in Eqn 4.5 and k_{df} in Eqn 4.13. The extinction coefficients for the total direct flux and the direct component of the direct flux are calculated from k_{df} with Eqns 4.13 and 4.14. A crop with a spherical leaf angle distribution can be taken as a reference. The theoretical value for k_{df}, assuming a scattering coefficient (σ) of 0.2 equals 0.72 for a canopy with a spherical leaf angle distribution (Goudriaan, 1977). However, actual values may substantially differ from this theoretical value. Firstly, the leaf angle distribution may not be spherical. A canopy with a planophile leaf area distribution has a higher extinction coefficient; a canopy with an erectophile leaf area distribution has a lower extinction coefficient than a canopy with a spherical leaf angle distribution. If the distribution of leaf area over leaf angle classes is known, k_{df} can be calculated using the procedure introduced by Goudriaan (1988). Otherwise k_{df} has to be measured.

Secondly, leaves are often clustered causing a lower extinction coefficient in the beginning of the growing season. The model accounts for these effects by using the measured value for k_{df}.

The extinction coefficient k_{df} has to be measured under an overcast sky. Direct radiation has to be avoided as the solar elevation determines the extinction coefficient for direct radiation (Eqn 4.13). If measurements have to be taken under a clear sky, a board can be used to shade the light measurement instrument. Light extinction can be measured by comparing radiation intensity above and below the canopy using a lightbar (generally a 1 m long tube with *PAR* light sensors built in). From the *LAI* and the measured light extinction, the extinction coefficient k_{df} can be calculated using Eqn 4.5. When global radiation is measured, k_{df} will be about 2/3 of the k calculated for global radiation, because absorption of near infrared radiation by the canopy is less efficient.

An important factor that may confound interpretation of measurements is the light absorption by organs other than leaves. This is accounted for in the model by specifically calculating light absorption by stems and storage organs. In calculating k_{df} of leaves from measurements, this effect should be accounted for. The extinction coefficient of erect leaves is used for stems and storage organs (0.4). The Specific Stem and Flower Areas (*SSA*, *SFA*) are calculated from measured area and dry weight. When measuring light extinction in the field, k_{df} of the leaves will be severely overestimated by light absorption of organs other than leaves if the light absorption is related to leaf area only. One option is to include the area of other organs in the calculation; the other option is to remove the flowers and then the leaves after the first light measurements and repeat the light measurements.

Values for k_{df} range from 0.40 - 0.70 for monocotyledons (erectophile) and 0.65 - 1.00 for dicotyledons (Monteith, 1969). For several species, estimates of k_{df} for photosynthetically active radiation (*PAR*, wavelength 400 - 700 nm) are presented in Table 7.1.

The k_{df} value of 0.69 for sugar beet was based on data from Clark and Loomis (1978). It fits with the value of 0.69 recorded by Tanaka (1983) and the value of 0.61 ± 0.04 reported by Kropff and Spitters (1992). For rice, a value of 0.40 is used early in the season to account for the vertical leaf angle distribution and clustering of leaves when transplanted on hills at 20 × 20 cm, and 0.60 later in the season when the canopy is closed and leaves became more droopy. For *C. album* the same value of k_{df} is used as for sugar beet (Kropff and Spitters, 1992). The value of 1.0 recorded by Monsi and Saeki (1953) is somewhat outside this range. This high value of Monsi and Saeki (1953) may be explained by light absorption by the flower stalks of *C. album* later in the season. At present, however, no estimate of their extinction characteristics is available. For wheat a value of 0.60 is used and for maize 0.65 (Spitters *et al.*, 1989a). For *E. crus-galli* a value of 0.80 or higher was determined from data in which the effect of stems and flowers was not taken into account.

CO_2 assimilation-light response of individual leaves

In the model, canopy CO_2 assimilation is calculated on the basis of the CO_2 assimilation-light response of individual leaves. This response follows a saturation type of function, characterized by the initial slope (the initial light use efficiency (ε, kg CO_2 ha^{-1} leaf h^{-1}/(J m^{-2} leaf s^{-1})) and the asymptote (A_m, kg CO_2 ha^{-1} leaf h^{-1}) (Eqn 4.20, Fig. 7.1).

For the initial light use efficiency (ε), a constant value of 0.45 - 0.50 kg CO_2 ha^{-1} leaf h^{-1} / (J m^{-2} leaf s^{-1}) is chosen, based on data of Ehleringer and Pearcy (1983). As they did not observe differences between species, such values are used for all species. In C_3 species, ε decreases slightly with increasing temperature as the affinity of the carboxylating enzyme Rubisco for O_2 increases compared to CO_2. In C_4 species, ε is independent of temperature as these plants have no photo-respiration (which is the reduction CO_2 assimilation by oxygenation of the carboxylating enzyme).

The light saturated rate of leaf CO_2 assimilation (A_m), however, varies considerably, mainly as a function of leaf age and the present environmental conditions and the environmental conditions to which the leaf has been exposed in the past. It is also influenced by genotype and plant species. A_m varies between 10 - 50 kg CO_2 ha^{-1} leaf h^{-1} for C_3 species and between 10 - 90 kg CO_2 ha^{-1} leaf h^{-1} for C_4 species, depending on leaf N concentration and temperature (Goudriaan, 1982; van Keulen and Seligman, 1987). An average value for an actively photosynthesizing leaf under favourable conditions is 40 kg

CO_2 ha^{-1} leaf h^{-1} for a C_3 species and 70 kg CO_2 ha^{-1} leaf h^{-1} for a C_4 species. The effect of temperature can be introduced with a species specific reduction function (Fig. 7.2). In the model these relationships are normalized with the maximum value set to 1.

For the relationship between leaf N concentration a linear relationship can be used (van Keulen and Seligman, 1987; Penning de Vries *et al.*, 1990; Sinclair and Horie, 1989). A good approximation is to relate A_m to N concentration expressed per unit leaf area because that determines the amount of chlorophyll per unit area. This relationship explains the decrease in A_m later in the growing season when the N concentration in the leaves decreases. For most species the effect of reduced N concentrations in the leaves during canopy senescence is accounted for by a reduction factor which is a function of developmental stage. For rice and *E. crus-galli* (in the Philippines) the N content of the leaves is input to the model. For rice, the maximum rate of leaf photosynthesis is calculated from the leaf N concentration using the relationship published by van Keulen and Seligman (1987): $A_m = -6.5 + 32.4N$, with N in g m^{-2} (Fig. 7.3). For *E. crus-galli*, the relationship given by Sinclair and Horie (1989) for maize

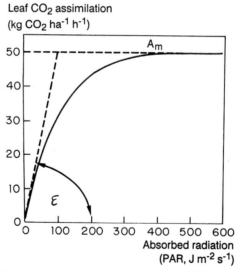

Fig. 7.1. The response curve of gross CO_2 assimilation of a single leaf versus the absorbed radiation (*PAR*). The tangent of initial slope is the initial light use efficiency (ε (kg CO_2 ha^{-1} leaf h^{-1})/(J m^{-2} leaf s^{-1})), and the asymptote is characterized by A_m (kg CO_2 ha^{-1} leaf h^{-1}).

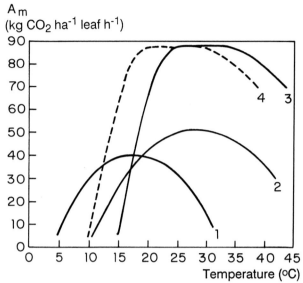

Fig. 7.2. Average relationship between maximum assimilation rate of single leaves at light saturation (A_m) and temperature for (1) C_3 crops from temperate climates, (2) C_3 crops from warm climates, (3) thermophile C_4 crops, and (4) cultivars of C_4 crops adapted to temperate climates. Redrawn after Versteeg and van Keulen (1986).

was used, as both species have the C_4 mechanism: $A_m = -27 + 135N$, with N in g m^{-2}.

Penning de Vries *et al.* (1989) compiled data from many sources on A_m for several crops. For maize and wheat values of 70 and 40 kg CO_2 ha^{-1} leaf h^{-1}, respectively, have been used; for *E. crus-galli* (in the Netherlands) 70 kg CO_2 ha^{-1} h^{-1} as this is a C_4 species like maize (Table 7.1).

The value of 50 kg CO_2 ha^{-1} leaf h^{-1} for sugar beet and *C. album* is within the wide range of values reported for sugar beet by, among others, Hall and Loomis (1972), Hodanova (1981) and van der Werf (Department of Theoretical Production Ecology, Wageningen Agricultural University, unpublished data), and for *C. album* by Chu *et al.* (1978), Pearcy *et al.* (1981), van Oorschot and van Leeuwen (1984) and our own data. No information was available on the CO_2 assimilation characteristics of the flower stalks of *C. album*.

Rate of net CO_2
assimilation (mg m^{-2} s^{-1})

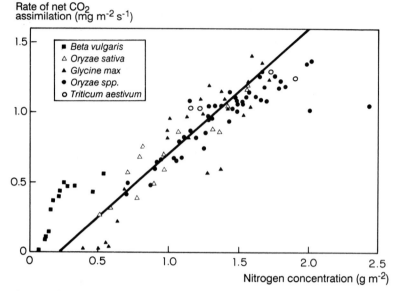

Fig. 7.3. The relation between the maximum rate of CO_2 assimilation of
single leaves and the leaf N concentration on a per area basis (g m^{-2}).
Redrawn after van Keulen and Seligman (1987).

Maintenance respiration

The maintenance requirements are more or less proportional to the
biomass to be maintained; typical values for the various plant organs
are given in Table 7.1. For leaves, stems and roots, in general, values
of 0.030, 0.015 and 0.010 kg CH_2O kg^{-1} dry matter d^{-1}, respectively,
are used. For rice a value of 0.020 kg CH_2O kg^{-1} dry matter d^{-1} was
used (Penning de Vries *et al.*, 1989). For storage organs the value can
be approached by calculating maintenance respiration for the active
tissue only, representing the envelope of the stored material like the
hull in rice, as the biomass stored is biochemically stable and does not
require maintenance. For some species like sugar beet, a species specific percentage of the biomass is inactive. For the fruits of *C. album*,
a value of 0.01 kg CH_2O kg^{-1} dry matter d^{-1} was assumed, based on
values for similar seed crops (Penning de Vries *et al.*, 1989). For sugar
beet, a storage component (sucrose) and a non-storage component are
distinguished. The sucrose is metabolically inactive and does not require maintenance, whereas the maintenance coefficient of the non-
storage component is assumed to be equal to that of the stem. With a
sugar content of 80% on a dry weight basis, this means a beet maintenance coefficient of $0.80 \times 0 + 0.20 \times 0.015 = 0.003$ kg CH_2O kg^{-1} dry

matter d^{-1}. The same value was used for rice. Maintenance requirements decrease with the metabolic activity of the plant. In the model, this is accounted for by assuming plant maintenance respiration to be proportional to the fraction of the accumulated leaf weight that is still green (Spitters *et al.*, 1989a). In this way, the maintenance cost for sugar beet at final harvest is reduced to approximately 60% of its potential value, as defined by the coefficients of Table 7.1. The actual maintenance coefficient of the storage beet as calculated in the model thus reduces to 0.0018 kg CH_2O kg^{-1} dry matter d^{-1}, a value well in line with those measured one or two days after harvest by, among others, Koster *et al.* (1980) (under the assumption of a reference temperature of 25 °C, $Q_{10} = 2$ and 24% dry matter in the beet).

Growth respiration

The primary assimilates in excess of the maintenance cost are converted into structural plant material. The amount of structural dry matter produced per unit of available carbohydrates depends on the chemical composition of the dry matter formed (Eqn 4.26). Typical values of the glucose requirements (Q) for various groups of compounds were derived on the basis of their chemical composition by Penning de Vries and van Laar (1982a, modified by Penning de Vries *et al.* (1989)) (Table 7.1). The value for the storage beet was derived from its composition as given by Penning de Vries *et al.* (1983), whilst that of *C. album* fruits was based on the measured chemical composition of the seeds, following the procedure described by Penning de Vries *et al.* (1989). A list of Q-values for storage organs of a large number of crop species is given by Penning de Vries *et al.* (1989).

Average light use efficiency

In the alternative approach to calculate dry matter production from absorbed radiation using the average light use efficiency, the detailed computations of the daily rates of CO_2 assimilation and respiration are replaced by a single equation in which daily dry matter increment is set proportional to intercepted light. So, this procedure is not included in the current version of the model INTERCOM. The average light use efficiency (E_{dm}, kg dry matter MJ^{-1}) is estimated by the slope of the linear regression of biomass on cumulative light interception. For that purpose, light interception can be calculated for each day during the season according to:

Table 7.1. Summary of the parameter estimates for sugar beet and *C. album*, maize, *E. crus-galli* (for the Netherlands and Philippines) and rice.

	Symbol	Unit	Value					
			Sugar beet	*C. album*	Maize	*E. crus-g.* (Neth.)	Rice	*E. crus-g.* (Phil.)
Light interception:								
extinction coefficient for PAR								
for leaves	k_{df}	m² m⁻²	0.69	0.69	0.65	0.80	0.40 0.60*	0.80
for stems	k_s	m² m⁻²	0.60	0.69	-	-	0.60	0.60
for flowers	k_f	m² m⁻²	0.60	0.69	-	-	0.60	0.60
Photosynthesis and respiration:								
maximum photosynthetic rate	A_m	kg CO₂ ha⁻¹ h⁻¹	50	50	70	70	Fig. 7.3	Fig. 7.3
initial light use efficiency	ε	kg CO₂ ha⁻¹ h⁻¹ (J m⁻² s⁻¹)⁻¹	0.50	0.50	0.45	0.45	0.45	0.45
average light use efficiency	E_{dm}	kg DM MJ⁻¹	0.0036	0.0040 (*t* < 900 °C d) 0.0025 (*t* > 900 °C d)	-	-	0.0028	-
Maintenance coefficients:		kg CH₂O kg⁻¹ DM d⁻¹						
leaves			0.030	0.030	0.030	0.030	0.020	0.030
stems			0.010	0.015	0.015	0.015	0.015	0.015
fibrous roots			0.010	0.010	0.010	0.010	0.010	0.010
storage organs			0.003	0.010	0.010	0.010	0.003	0.010

Table 7.1. Continued

	Symbol	Units						
CH$_2$O requirements	Q	kg CH$_2$O kg^{-1} DM d^{-1}						
leaves			1.46	1.56	1.47	1.47	1.326	1.47
stems			1.51	1.51	1.52	1.52	1.326	1.52
fibrous roots			1.44	1.44	1.45	1.45	1.326	1.45
storage organs			1.29	1.49	1.49	1.49	1.462	1.49
Phenology:								
base temp. for development	T_b	°C	2	2	10	10	8	10
max. temp. for development	T_m	°C	21	-	-	-	24	24
Dry matter distribution pattern:								
partitioning coefficients	pc	-	Fig. 7.6					
Leaf area dynamics:								
the initial leaf area per plant	$L_{p,0}$	cm^2 plant^{-1}	0.45	0.13	6.69	0.368	17**	0.368
base temp. for leaf development		°C	3	3	10	10	8	10
relative growth rate of leaf area	R_l	°C^{-1} d^{-1} m^{-2} leaf	0.0158	0.0186	-	-	0.0085	0.012
		m^2 ground DS^{-1}	-	-	12.5	11.4	-	-
Specific Leaf Area	SLA	m^2 leaf kg^{-1} leaf	Fig. 7.8				Fig.7.12	Fig. 7.12
leaf senescence parameter	R_s		-0.00055	-0.0022	-	-		
Plant height:								
height increment parameters	h_m	m	0.60	1.60	2.05	1.01	0.74	1.32
	b	-	67	298	3.49	4.05	10.24	12
	s	°C^{-1} d^{-1}	0.007	0.009	-	-	0.0042	0.0031
		DS^{-1}	-	-	6.54	6.29	-	-

* When LAI > 1   ** For 21-day-old seedlings (3 plants per hill).

$$I_a = I_{0,d} (1 - \rho) (1 - \exp(-k_{df} LAI)) \qquad (7.1)$$

where the *LAI* is obtained by nonlinear interpolation between its value measured at the two nearest sampling dates, and *PAR* is estimated as 50% of the total solar irradiance measured at a nearby weather station. Daily amounts of intercepted light have to be accumulated after plant emergence and plotted against total biomass production (Fig. 7.4). For sugar beet and *C. album* data from field experiments conducted in 1984, 1985 and 1986 were used (Chapter 8;

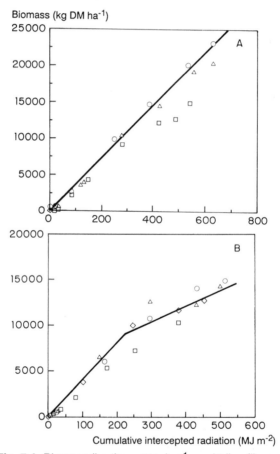

Fig. 7.4. Biomass (kg dry matter ha^{-1}; excluding fibrous roots) in relation to cumulative light interception for (A) sugar beet (1984 □ ; 1985 o and 1986 △) and (B) *C. album* (1984 □ ; 1985 o; 1986a △ and 1986b ◊) grown in monocultures. Drawn lines are estimates of relationships at ample water supply.

Kropff and Spitters, 1992). For the 1985 data for sugar beet, the relationship between biomass production and cumulative light interception fitted rather well to a straight line, indicating a more or less constant value of E_{dm} throughout the season. In the 1984 and 1986 experiments, however, the relationship bent off later in the season due to water shortage. The value of E_{dm} at ample water supply is input for the summary model. The estimated efficiency of 0.0036 kg dry matter MJ^{-1} (Table 7.1) agrees with values reported by Milford and Riley (1980), if allowance is made for the fact that they considered total radiation instead of *PAR*. In Fig. 7.4B, a distinct bend emerged for *C. album*, not only in the 1984 and 1986 data where it may partly be ascribed to water shortage, but also in the 1985 data when water supply was optimal. This bend occurred around 900 °C d after plant emergence and corresponded to the time that the fruit bearing stalks started to intercept a large part of the incident light. Table 7.1 gives the estimated parameter values. For rice a value of 0.0028 was reported by Sinclair and Horie (1989).

Phenology

For sugar beet and *C. album*, the phenological development was described using the temperature sum approach, because sugar beet does not flower or mature in the (first) growing season (Eqn 4.29). For other crops the scale for development stage (*D* or *DVS*) was used (0 is emergence, 1 is flowering, 2 is maturity) (Eqn 4.30). The temperature sum is calculated on the basis of daily average temperatures. For sugar beet, a base temperature of 2 °C was used, below which no development occurs and a maximum temperature of 21 °C, above which the development rate is not further accelerated by increasing temperature. These cardinal points were inferred from temperature responses of leaf appearance rate (Clark and Loomis, 1978; Hodanova, 1981; Milford and Riley, 1980; Terry, 1968).

For *C. album*, also a base temperature of 2 °C was assumed. This choice was inspired by the narrow range in base temperatures of 0 °C to 3 °C encountered by Angus *et al.* (1981) for species originating from temperate regions. In the experiments presented, flowering of *C. album* started about 500 °C d after seedling emergence. To account for photoperiodic effects on flowering in *C. album* (Ramakrishnan and Kapoor, 1973; Warwick and Marriage, 1982) as well as different temperature requirements between ecotypes, the development stage has to be rescaled to a dimensionless variable having the value 0 at seedling emergence, 1 at the onset of flowering, and 2 at maturity. The effect of photoperiod could be included by using one of the procedures mentioned in Chapter 4 (Section Phenological development).

For rice, maize and *E. crus-galli* the dimensionless scale for development was used (Eqn 4.30). The development rate (d^{-1} or $({}^{\circ}C\ d)^{-1}$) was calculated from experimental data by the inverse of the number of degree-days between two phenological events. A Fortran program for PCs calculates these parameter values from field observations (Program DRATES, available upon request, M.J. Kropff, International Rice Research Institute, Los Banos, Philippines).

Dry matter partitioning

In the model, the total daily dry matter increment is partitioned to the various plant organ groups according to factors that are a function of the development stage; the development stage being expressed in ${}^{\circ}C$ d after plant emergence or in the dimensionless variable *DVS*. These factors are derived by analysing the fractions of new dry matter production allocated to the plant organs between two subsequent harvests. The partitioning coefficient (pc_k) for organ k can be calculated by dividing the change in weight of organ k (ΔW_k) between two samplings by the total change in weights of the organs ($\Sigma \Delta W_j$):

$$pc_k = \frac{\Delta W_k}{\Sigma \Delta W_j}\ ; \qquad \Delta W_k \geq 0 \tag{7.2}$$

If the change in weight of a specific organ is negative, no growth is assumed. These losses are accounted for separately and are thus not included in the partitioning functions.

The partitioning coefficients are calculated for all sampling intervals. The development stage at the middle of the period between two samplings can be calculated with the program DRATES to obtain the partitioning coefficients as a function of development stage (or temperature sum).

The relationships used in the model are given in Figs 7.5 and 7.6 for the six species. The total daily produced dry matter is first partitioned between 'shoot' (for sugar beet including beet tap root) and 'root' (Fig. 7.5). The partitioning between the different 'shoot' organs is represented in Fig. 7.6. The dry matter distribution patterns are based on several series of experiments (cf. Kropff and Spitters, 1992).

Redistribution of assimilates that have been stored in the stems is accounted for in a species specific manner. For rice and *E. crus-galli*, the measured reduction in stem weight, defined as the difference between the maximum stem weight and the final stem weight, is

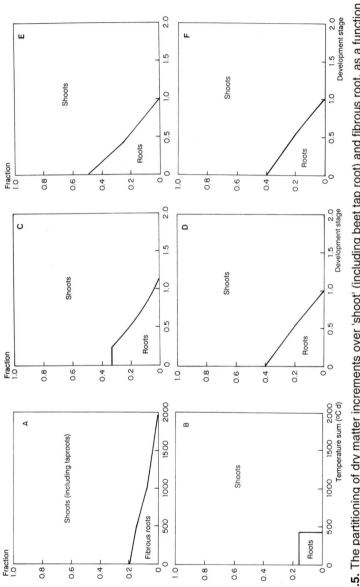

Fig. 7.5. The partitioning of dry matter increments over 'shoot' (including beet tap root) and fibrous root, as a function of temperature sum after emergence or development stage. (A) Sugar beet: inferred from data of Boonstra (unpublished); (B) *C. album*: guesstimate, partly based on the interrelationship of root and leaf growth; (C) maize and (D) *E. crus-galli*: data from Akita (IRRI, cv. IR64, unpublished); (E) rice: data from Kropff *et al.* (1984); (F) same as (D).

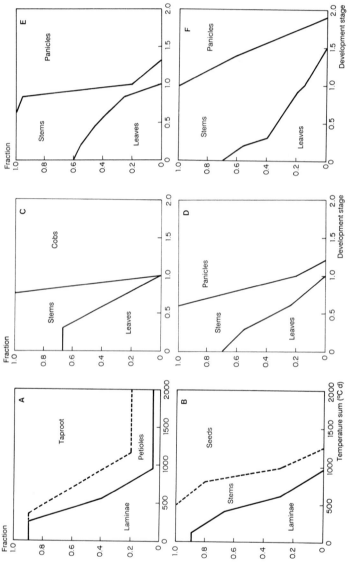

Fig. 7.6. The partitioning of 'shoot' dry matter increments among the various 'shoot' organs as a function of the temperature sum after emergence for sugar beet (A) and *C. album* (B), and as a function of development stage for (C) maize and (D) *E. crus-galli*: data from Kropff *et al.* (1984); (E) rice: data for cv. IR72 (IRRI 1991 wet season, unpublished); (F) *E. crus-galli*: estimated data from IRRI, 1992 dry season (unpublished).

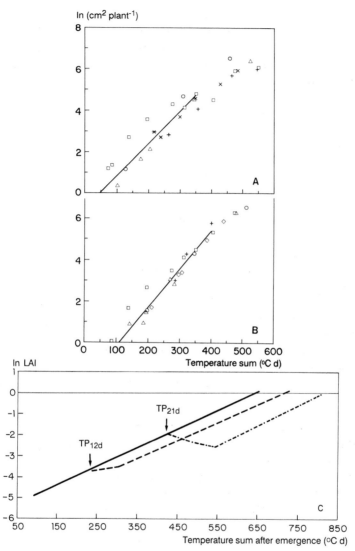

Fig. 7.7. Relationship between the logarithm of leaf area per plant (*L*) and the temperature sum (*ts*) after plant emergence for sugar beet (A) and *C. album* (B). Estimated relationships:

Sugar beet: $\ln L_{ts} = -0.79 + 0.0158ts$ $ts \leq 350\ ^{\circ}C\ d$ $(\overline{T}=12.7\ ^{\circ}C)$
C. album: $\ln L_{ts} = -2.05 + 0.0186ts$ $ts \leq 400\ ^{\circ}C\ d$ $(\overline{T}=14.3\ ^{\circ}C)$

where $\overline{T}$ is the observed average temperature during the period considered. Symbols as in Fig. 7.4; extra symbols refer to unpublished data from CABO-DLO, Wageningen. (C) For rice (IRRI, 1991 wet season) the relationship was determined for direct-seeded (solid line), transplanted after 12 days (dashed line) and transplanted after 21 days (dot-dashed line).

assumed to be the amount of stem reserves that are translocated to the grains. The reserves are converted into CH_2O and added to the amount of assimilates available for growth. For other species, weight fractions of stems and leaves that can be translocated have been estimated and are introduced as a fraction of the death rate of the organs.

Leaf area

In the early phases, leaf area growth proceeds more or less exponentially, the relative growth rate being approximately linearly related to temperature (Eqn 4.33). When leaf area per plant is plotted on a logarithmic scale against the temperature sum after emergence, a more or less linear relationship is, therefore, obtained (Fig. 7.7). The slope measures the relative leaf area growth rate (R_l, $°C^{-1}$ d^{-1}) and the intercept the apparent leaf area at emergence. *C. album* had a smaller intercept ($L_{p,0}$, the initial leaf area per plant at seedling emergence (m^2 $plant^{-1}$)) than sugar beet because of smaller seed reserves. *C. album* showed a slightly greater R_l than sugar beet, mainly explained by a greater part of the assimilates allocated to the leaf blades (Fig. 7.6B). The exponential phase ended when an increasing portion of the assimilates was allocated to non-leaf tissue (Fig. 7.6B). Fig. 7.7C gives the relationships found in direct-seeded and transplanted rice, showing that the transplanting shock only causes a delay, but does not affect the slope. For nitrogen limiting situations, the relative growth rate of the leaf area has to be made a function of leaf N content. Miyasaka *et al.* (1975) reported a linear relation between the relative growth rate of the leaf area in rice and the leaf N content.

After this exponential phase leaf area growth is simulated by multiplying the leaf dry weight increment by the specific leaf area (*SLA*, m^2 leaf kg^{-1} leaf)) of the newly formed leaves (Eqn 4.36). *SLA* is plotted in Fig. 7.8 as a function of the development stage expressed in degree-days for the six species. For each harvest interval, the *SLA* of new leaves was calculated by dividing the increase in leaf area index (*GLAI*) by the increase in leaf weight (G_{lv}) between subsequent harvests. In assessing the relationships in Fig. 7.8, the scatter in the data points was accommodated by also considering the relationship of LAI/W_{lv} (weight of the leaves) which gives more stable figures at later developmental stages.

In calculating *SLA* of sugar beet, the midrib was excluded from the leaf blade from five weeks (approximately 300 °C d, $T \geq 2$ °C) after plant emergence onwards. The decrease in leaf area due to senescence was estimated from Fig. 7.9, where the green leaf area index is

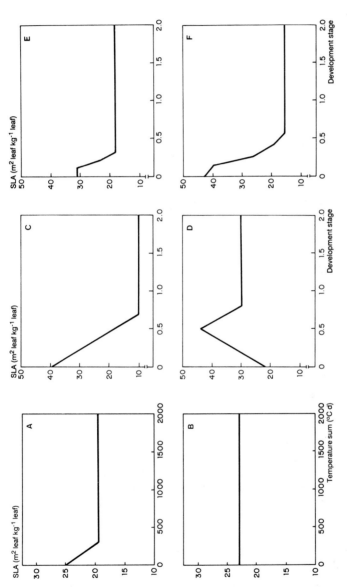

Fig. 7.8. The specific leaf area of the newly formed leaves as a function of the temperature sum after emergence for sugar beet (A) and *C. album* (B). The constant levels are 19.4 and 23.1 (m² leaf kg⁻¹ leaf) for sugar beet and *C. album*, resp. *SLA* as a function of development stage for (C) maize and (D) *E. crus-galli*: data from Kropff et al. (1989); (E) rice: data from Penning de Vries *et al.* (1984); (F) *E. crus-galli*: data from Kropff *et al.* (IRRI, unpublished).

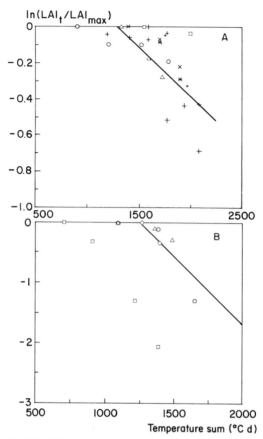

Fig. 7.9. Time course of the natural logarithm of green leaf area index (*LAI*) relative to its maximum value (*LAI*$_{max}$) for sugar beet (A) and *C. album* (B). Time as temperature sum in °C d after plant emergence. Symbols as in Fig. 7.4; additional data for sugar beet from unpublished data of Sibma (CABO, Wageningen). Slope of the drawn lines gives value of the relative death rate of leaves (*R*$_s$) used in the model. Source: Kropff and Spitters, 1992.

depicted relative to its maximum value in the experiment concerned (Eqn 4.37). This procedure assumes that during the senescence phase there is no increment of new leaf area at all. This holds for *C. album*. In sugar beet, however, this assumption is not fully true, which biases the estimation of the relative death rate. However, this hardly disturbs the simulation results of sugar beet yields because leaf senescence occurred not before late in the season. In *C. album*, leaf senescence appeared much earlier in the 1984 experiment than in both the other experiments (Fig. 7.9), probably due to drought in this experiment. The 1984 data were, therefore, excluded in establishing the senescence function. Similar procedures were followed for the other species to estimate these parameters (Table 7.1).

For rice and *E. crus-galli* (in the Philippines) a different procedure was followed. A relative death rate was defined based on leaf weight instead of leaf area as a function of development stage (Fig. 7.10). The change in *LAI* is then calculated through the *SLA*.

Relative death rate of the leaves

The relative death rate of the leaves (*RDR*, d^{-1}) was calculated in a simplified way from experimental data. For the time interval between two samplings the relative death rate can be calculated as follows, starting at the time where the leaf dry matter is highest:

$$RDR = (\ln W_t - \ln W_{t+dt}) / dt \qquad (7.3)$$

in which *t* is expressed in days. Using the developmental rate program (DRATES), the development stages at the sampling dates are calculated. To relate the relative death rate to the development stage, the calculated relative death rate is assumed to be the rate at the average development stage between the samplings (Fig. 7.10).

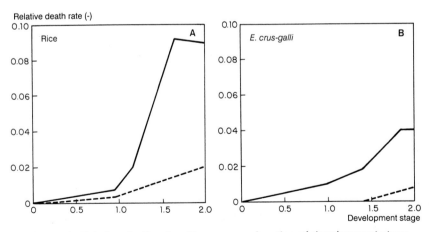

Fig. 7.10. Relative death rate of leaves as a function of development stage derived for rice and *E. crus-galli* (1992 Dry Season experiment). Source: Kropff *et al.*, unpublished.

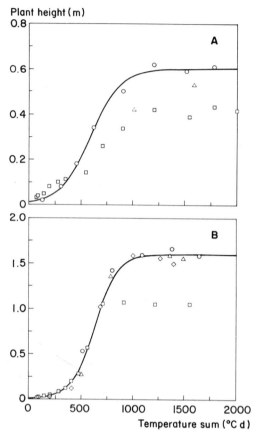

Fig. 7.11. Plant height (h) as a function of temperature sum (ts) after plant emergence, fitted by the logistic equation $h_{ts} = h_m/ (1 + b \cdot \exp(-s \cdot ts))$.
(A) Sugar beet: $h_m = 0.60$ m, $b = 67$, $s = 0.007$ (°C d)$^{-1}$
(B) *C. album*: $h_m = 1.60$ m, $b = 298$, $s = 0.009$ (°C d)$^{-1}$
Symbols as in Fig. 7.4. Source: Kropff and Spitters, 1992.

Plant height

The time course of plant height is described by a logistic function of the temperature sum after plant emergence (Eqn 4.39). The parameters s, b and h_m can easily be estimated by linearization of Eqn 4.39:

$$\ln ((h_m - h) / h) = \ln b - s \cdot ts \qquad (7.4)$$

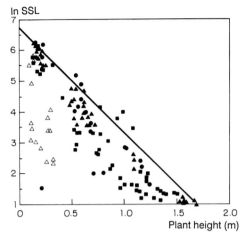

Fig. 7.12. The relation between the Specific Stem Length (*SSL*) and plant height for *C. album*, measured in different experiments in monoculture and competition situations. Symbols indicate different growth stages (Δ: 0 < *ts* < 500; ■: 600 < *ts* < 800; ●: 100 < *ts* < 1300; ▲: 1300 < *ts* < 1500). The line indicates the maximum *SSL* possible in relationship to height (Eqn 4.41).

Firstly, h_m has to be estimated from the asymptote (Fig. 7.11). To account for limitations in height growth as a result of competition, Eqn 4.41 can be used which defines the maximum specific stem length as a function of plant height. Until now this relationship has only been evaluated for *C. album* as in the series of experiments with sugar beet and *C. album* strong competition effects on height of *C. album* were observed. The relationship will be different for other species. Fig. 7.12 gives the relationship between the maximum specific stem length and plant height. It was compiled using data from extreme situations when *C. album* emerged early and late in a sugar beet crop. It shows that *C. album* has a high phenotypic plasticity with respect to height growth. The later data of the 1984 experiment were excluded because plant height growth was reduced by drought stress, an effect which is accounted for in the model.

Plant-water relationships

Only a few species specific parameters determine variation in plant-water relationships in the model. The maximum rooted depth is an important factor, although in the simple water balance used in the model INTERCOM, the effect of differences in rooted depth between competing species is not accounted for. The crop factor (Eqn 5.13) is set to 0.9 for C_3 species (Feddes, 1987) and 0.7 and 0.9 for C_4 species in temperate and tropical climates, respectively. Also the soil depletion factor (Eqn 5.15) is different for C_3 and C_4 species (Doorenbos and Kassam, 1979, Fig. 5.4). Because the nitrogen balance has not been included in the model INTERCOM so far, we refer to Chapter 6 for details on parameters.

Discussion and conclusions

An overview of the most important parameters to characterize morphological and eco-physiological traits is given in Table 7.1. Major differences between species are found for the light extinction coefficient, photosynthetic characteristics (C_3 - C_4 difference), phenological parameters, leaf area dynamics and plant height. In Chapter 8, the relative importance of these characteristics with respect to competition will be evaluated.

Chapter Eight

Understanding Crop-Weed Interaction in Field Situations

M.J. Kropff, S.E. Weaver, L.A.P. Lotz, J.L. Lindquist, W. Joenje, B.J. Schnieders, N.C. van Keulen, T.R. Migo and F.F. Fajardo

Introduction

Processes related to competition for the different resources light, water, and nitrogen were discussed separately in Chapters 4 to 7. However, to understand the competition process in agricultural systems, these aspects have to be integrated as has been done in the model INTERCOM and the total model has to be evaluated with respect to its capacity to explain and predict phenomena observed in the real system through field experimentation. It is not until model performance has been thoroughly evaluated that the model can become useful for applications like explaining backgrounds of differences in yield loss between treatments or determining which factors are most critical in determining yield loss by weeds or in the design of competitive plant types for plant breeders.

The process of model evaluation is often also indicated by the terms validation and verification. Thornley and Johnson (1990) indicated that the term validation is in principle not correct because a theory or model can only be falsified. As we deal with a research model here, we would like to see if the model is capable of explaining competitive phenomena observed in field experiments on the basis of the mechanisms included in the model. In contrast to predictive models, where it is not a problem if the assumptions are not sound as long as the model predicts accurately, it is important for application of this model that it is as mechanistic as possible. If, for example, the model is being used for designing plant types with high competitive ability, wrong assumptions in the model will lead to wrong conclusions and research directions.

Therefore, it is important not only to test the final output of the model, but evaluate components of the model as well. The general approach we followed is a first evaluation of the monoculture models for both the crop and the weeds. For some species characteristics of the weeds, the only data available to derive parameter values were data from the field experiments that were designed for the evaluation of the competition model. For parameterization, however, we only used data from the monoculture treatments (Chapter 7). Wherever possible, multiple data sets were used for model evaluation covering a wide range of competition situations. In the analyses, a single set of parameters for species characteristics was used to explain the variation in yield loss between experiments or treatments observed.

In this chapter, we will first shortly describe the history of the evaluation process of the model INTERCOM which was strongly intertwined with model development. This was an iterative process, in which model evaluation led to changes in underlying concepts. Then two case studies one with rice and *Echinochloa crus-galli* L. and one with sugar beet with *Chenopodium album* L. will be discussed in detail. Studies with other species will be discussed only briefly.

History of the development and evaluation of the model INTERCOM

The first version of the eco-physiological competition model was tested with data from competition experiments with maize, *Sinapis arvensis* L. and *Echinochloa crus-galli* L. (Kropff *et al.*, 1984; Spitters and Aerts, 1983). This model was based on the models for monoculture crops BACROS (de Wit *et al.*, 1978), SUCROS (van Keulen *et al.*, 1982) and the model for water-limited production described by van Keulen (1975). These eco-physiological models for monocultures have been evaluated and tested for several monoculture crops (de Wit *et al.*, 1978; van Keulen, 1975; van Keulen and Seligman, 1987; Penning de Vries and van Laar, 1982a; Rabbinge *et al.*, 1989; Kropff, 1990). Components of these models were also evaluated, like the submodel for canopy photosynthesis, which was validated by using experimental data from enclosure studies (van Keulen and Louwerse, 1975; Kropff and Goudriaan, 1989). Fig. 8.1 shows the performance of the routines for canopy photosynthesis for a cloudy and a clear day for a Faba bean crop (Kropff and Goudriaan, 1989). The performance of the model INTERCOM for monocultures is indicated in Table 8.1.

The model INTERCOM was developed from the first version of the competition model through gradual improvement of parts of the model. That was needed because the first preliminary version was very complex, too sensitive to parameter values and needed calibration per experiment (e.g. the site factor used by Spitters, 1984). A major improvement was the introduction of a new procedure to simulate leaf area development (Kropff, 1988c). The original model

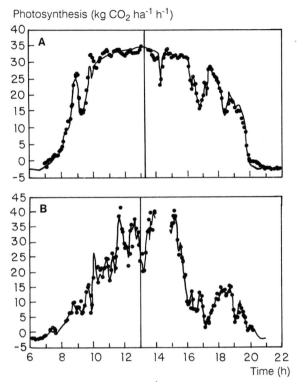

Fig. 8.1. Simulated (solid line) and observed (•) diurnal canopy photosynthesis (kg CO_2 ha^{-1} h^{-1}) on (A) a clear and (B) a cloudy day. Source: Kropff and Goudriaan, 1989.

simulated leaf area from dry matter increment using the specific leaf area (*SLA*) from emergence onwards, assuming source-limited leaf area development throughout the growing season. This caused an extreme sensitivity in the model for all parameters including CO_2 assimilation related parameters: small differences in parameter values had strong implications for leaf area development. This was the result of the positive feedback during the exponential growth phase (more leaf area results in higher growth rates, which results in more leaf area on the next day). This sensitivity was strongly reduced by introducing the new procedure for leaf area development as discussed in Chapter 4. For example, in a simulated mixture of two identical maize varieties, that had equal yields in mixtures, a reduction of the value of the Specific Leaf Area of one of the competing varieties by 20%, resulted in a five times lower yield compared to the other variety. If the same was done for photosynthesis related parameters (A_m and ε) a two to three fold lower yield was simulated. In the current model, a 5% change in one of these parameters resulted in only a 1 - 10% change in yield loss, depending on the competition situation (see Chapter 9). Other major improvements were the simulation procedure for canopy photosynthesis and simplifications for the calculation of transpiration, evaporation and the water balance of the soil (Kropff, 1988a).

The eco-physiological model INTERCOM was used to analyse five field experiments on sugar beet and *C. album* competition (Kropff, 1988a; Kropff *et al.*, 1992b) and maize - *E. crus-galli* (Kropff *et al.*, 1992a). Additional validation of the eco-physiological model was performed using independent data from critical period experiments with the same species (Weaver *et al.*, 1992). Subsequently, the model was used to analyse data on tomato - *Amaranthus retroflexus* L. and tomato - *Solanum ptycanthum* Dun. competition (Kropff *et al.*, 1992a; Weaver *et al.*, 1992) in Canada. Lotz *et al.* (1990) analysed experimental data on the effect of omitting herbicide applications in winter wheat in the Netherlands using the model. Recently, the model was used to analyse rice - *E. crus-galli* competition situations (Kropff and Lindquist, in prep.).

Table 8.1. Observed and simulated yields of weed-free crops, in Harrow (Canada) and Wageningen (Netherlands) (Kropff *et al.*, 1992a); rice in Los Banos (Philippines). Source: Kropff *et al.*, unpublished.*

Crop	Site	Year	Observed yield** (kg ha^{-1})		Simulated yield (kg ha^{-1})
Tomato (seeded)	Harrow	1984	3172 $\pm$	222	3009
Tomato (seeded)	Harrow	1985	2704 $\pm$	260	3290
Tomato (transpl.)	Harrow	1984	2736 $\pm$	164	2990
Tomato (transpl.)	Harrow	1985	4189 $\pm$	330	4312
Maize	Wageningen	1982	13110 $\pm$	1940	13901
Maize	Wageningen	1983	8440 $\pm$	210	8459
Sugar beet	Wageningen	1984	14900 $\pm$	1397	14870
Sugar beet	Wageningen	1985	23100 $\pm$	1233	20644
Sugar beet	Wageningen	1986	20400 $\pm$	687	20450
Rice	Los Banos	1992	7068 $\pm$	169	7002

* Yields of tomato, maize, sugar beet and rice represent fruit, grain, root and panicle dry weight, respectively.

** Means $\pm$ standard errors.

Case study 1: Rice and *E. crus-galli*

Echinochloa species are among the most severe weed species causing problems in rice crops (Lubigan and Vega, 1971; Noda, 1973; Smith, 1983; Rao and Moody, 1987). In transplanted rice, seedlings are often transplanted with the rice seedlings because of their morphological similarities (Rao and Moody, 1987). That reduces the advantage of transplanted rice plants over newly emerged weeds, resulting in tremendous yield losses which are similar to the losses observed in direct seeded rice (Rao and Moody, 1992).

The model INTERCOM was evaluated first using data from an experiment with irrigated direct seeded rice and *E. crus-galli*, conducted by Kropff, Moody, Migo and Fajardo (unpublished data) in the 1992 dry season at the IRRI experimental farm in Los Banos, Philippines. The experiment consisted of two weed density treatments combined in a factorial with two weed emergence dates in four replicates. In the early emergence treatment, the weeds emerged two days after the crop at 100 and 56 plants m^{-2}, and in the late emergence treatment, the weeds emerged 22 days after the crop at 19 and 13 plants m^{-2}. Monocultures of both weed and crop at a single density were included. Pregerminated rice and *E. crus-galli* seeds were broadcast in saturated soil. Destructive plant sampling was done periodically throughout the growing season. At each harvest date, samples were separated into green leaves, stems, and panicles. Measurements taken included leaf area, nitrogen concentration in green leaves, and plant height.

The parameter values for species characteristics as reported in Chapter 7 were used for all simulations. Rice and *E. crus-galli* dry matter production and leaf area development were simulated accurately, using the measured specific leaf N (g N m^{-2}) as input in the model (Fig. 8.2). Using the same set of parameters, good simulations were obtained for the mixtures (Fig. 8.2). In these experiments, yield loss was moderately low because of the low relative weed density (700 rice plants m^{-2} versus 54 *E. crus-galli* plants m^{-2} for the early emerging treatment and 13 plants m^{-2} for the late emerging weeds).

Subsequently, the model was evaluated using a wide range of data sets on competition between *Echinochloa* species and transplanted or direct-seeded rice, which were summarized by Hill *et al.* (1990). Data sets were from by Noda *et al.* (1968), Smith (1968), Lubigan and Vega (1971), Hill *et al.* (1990), Rao and Moody (1992), and Kropff *et al.* (unpublished). Information on crop and weed densities and yield loss is given in Table 8.2. Input in the model was the date of emergence of the crop and the weeds, crop and weed density

Chapter 8

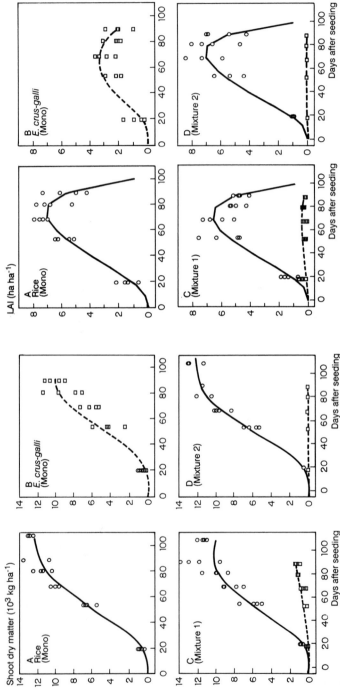

Fig. 8.2. Simulated (rice: solid line; *E. crus-galli*: dashed line) and observed (rice: o; *E. crus-galli*: □) total shoot dry matter, leaf area index, and weight of green leaves and weight of panicles for a competition experiment with direct-seeded rice and *E. crus-galli* in 1992, Los Banos, Philippines. Treatments: (A) monoculture rice, (B) monoculture *E. crus-galli*, (C) mixture of 700 rice plants m^{-2} and 54 *E. crus-galli* plants m^{-2}, that emerged two days after the crop, and (D) mixture of 700 rice plants m^{-2} and 13 *E. crus-galli* plants m^{-2} that emerged 22 days after the crop.

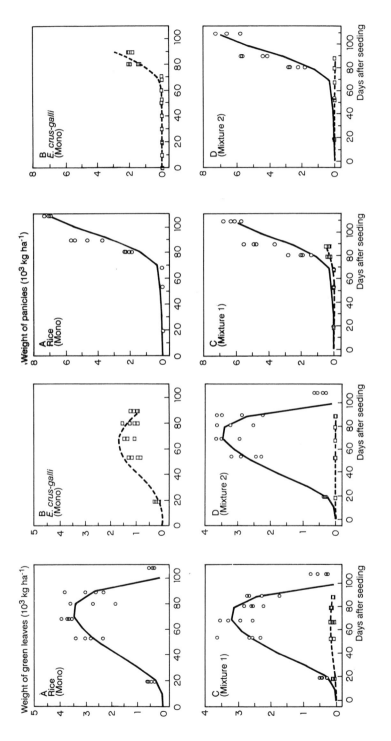

Fig. 8.2. Continued.

Chapter 8

Table 8.2. Summary of relative yields, crop and weed densities, year and cultivation method (DS = direct-seeded, T = transplanted) for experiments used for model evaluation.

Location	Cul.	Year	Weed dens. (pl. m⁻²)	Rice dens. (pl. m⁻²)	Rel. yield (%)	Location	Cul.	Year	Weed dens. (pl. m⁻²)	Rice dens. (pl. m⁻²)	Rel. yield (%)
Noda *et al.* (1968)	T	1965	0	11	100	Hill *et al.* (1990)	DS	1981	0	256	100
Kyushu, Japan	T	1965	5	11	91	California, USA	DS	1981	14	256	83
	T	1965	10	11	61		DS	1981	27	256	65
	T	1965	20	11	60		DS	1981	54	256	43
	T	1965	40	11	47		DS	1981	108	256	23
	T	1965	80	11	24		DS	1981	0	318	100
	T	1965	160	11	17		DS	1981	19	318	67
	T	1965	320	11	6		DS	1981	38	318	61
	T	1965	0	16	100		DS	1981	76	318	44
	T	1965	5	16	87		DS	1981	152	318	27
	T	1965	10	16	75		DS	1982	0	291	100
	T	1965	20	16	61		DS	1982	8	291	93
	T	1965	40	16	66		DS	1982	15	291	87
	T	1965	80	16	44		DS	1982	30	291	80
	T	1965	160	16	29		DS	1985	0	291	100
	T	1965	320	16	17		DS	1985	34	291	73
	T	1965	0	25	100		DS	1985	67	291	64
	T	1965	5	25	95		DS	1985	134	291	49
	T	1965	10	25	92		DS	1985	269	291	35
	T	1965	20	25	82		DS	1981	0	254	100
	T	1965	40	25	72		DS	1981	19	254	70
	T	1965	80	25	59		DS	1981	38	254	61
	T	1965	160	25	27		DS	1981	76	254	43
	T	1965	320	25	22		DS	1981	152	254	36
	T	1966	0	16	100		DS	1982	0	259	100
	T	1966	2	16	99		DS	1982	8	259	88
	T	1966	5	16	83		DS	1982	15	259	80
	T	1966	10	16	85		DS	1982	30	259	74
	T	1966	20	16	68	Kropff *et al.*	T	1991	0	125	100.0
	T	1966	40	16	59	(unpublished)	T	1991	24	125	78.9
	T	1966	80	16	49	Los Banos,	T	1991	30	125	74.0
	T	1966	160	16	37	Philippines	T	1991	35	125	76.3
	T	1966	320	16	23		T	1991	46	125	75.0
Lubigan and Vega	T	1969	0	20	100		T	1991	0	125	100.0
(1971)	T	1969	20	20	64		T	1991	10	125	102.1
Los Banos,	T	1969	40	20	40		T	1991	13	125	97.1
Philippines	T	1969	60	20	36		T	1991	13	125	105.9
	T	1969	80	20	24		T	1991	40	125	106.1
	T	1969	100	20	20	Rao and Moody	T	1989	0	90	100.0
Smith (1968)	DS	1962-3	0	32	100	(1992)	T	1989	30	60	10.0
Arkansas, USA	DS	1962-3	11	32	43	Los Banos,	DS	1989	0	90	100.0
	DS	1962-3	54	32	20	Philippines	DS	1989	30	60	5.9
	DS	1962-3	269	32	5	Kropff *et al.*	DS	1992	0	700	100
	DS	1962-3	0	107	100	(unpublished)	DS	1992	178	0	-
	DS	1962-3	11	107	60	Los Banos,	DS	1992	56	700	88.3
	DS	1962-3	54	107	34	Philippines	DS	1992	100	700	84.0
	DS	1962-3	269	107	11		DS	1992	13	700	93.0
	DS	1962-3	0	334	100		DS	1992	19	700	92.8
	DS	1962-3	11	334	75						
	DS	1962-3	54	334	51						
	DS	1962-3	269	334	21						

and weather data. For the data sets of Kropff *et al.* (unpublished), 1991 wet season (WS) and the 1992 dry season (DS), actual weather data were available. For the other experiments, actual weather data were not available. So standard weather data were used (Los Banos, 1992 dry season). Hill *et al.* (1990) already demonstrated using regression models that the differences between the data of Smith (1968), Hill *et al.* (1990), Noda *et al.* (1968) and Lubigan and Vega (1971) could be fully explained by differences in relative weed density (dates of emergence of crop and weed were the same). So, for these experiments yield loss differences are only explained by the model based on differences in crop and weed density and establishment method (transplanted or direct-seeded). Because crop and weeds generally emerged simultaneously in the experiments, no strong impact of weather on the relative competition effects are to be expected (Kropff *et al.*, 1992a). For example, the rice - *E. crus-galli* model predicted a yield loss of 22% in the dry season at 50 weed plants m^{-2} whereas 19% yield loss was simulated in the wet season at yield levels of 6600 kg ha^{-1} and 5200 kg ha^{-1}, respectively.

Some species characteristics could be made more specific using observations from the experiments. Height growth of rice and *E. glabrescens* in the experiment of Rao and Moody (1992) was reported and differed from the values we derived for *E. crus-galli*. Therefore, height growth parameters based on their data were used in the simulations of their experiments.

Simulated yield loss was predicted very accurately by the model over this wide range of competition situations (Fig. 8.3). These simulations confirm the conclusion of Hill *et al.* (1990) that the relative weed density and establishment method (transplanted or direct-seeded) were the major factors causing differences among treatments and experiments. Fig. 8.3 shows the low competitive ability of direct-seeded rice compared to transplanted rice. The relation between observed and simulated yield loss for all experiments is given in Fig. 8.4. The model explained 93% of the variation and the regression of simulated versus observed is close to the 1:1 relationship ($y = -0.6 + 0.99x$) (Fig. 8.4). This means that the differences in yield loss between the experiments can be explained by the eco-physiological model based on crop density, weed density and the period between crop and weed emergence and establishment method of rice.

Case study 2: Sugar beet and *C. album*

A very detailed study was conducted with sugar beet and *C. album*. In this study, model development and experimentation were

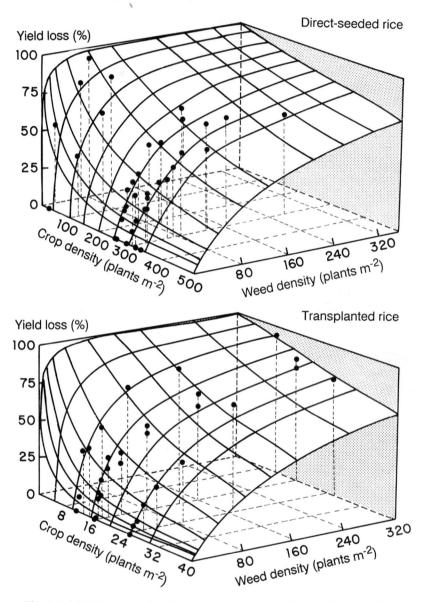

Fig. 8.3. Yield loss as a function of crop-weed densities as simulated by
the model INTERCOM (drawn response surface) and observed data (•)
for transplanted rice (data from Noda *et al.*, 1968; Lubigan and Vega,
1971), and for direct-seeded rice (data from Smith, 1968; Hill *et al.*,
1990).

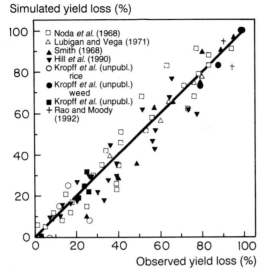

Fig. 8.4. The relationship between simulated and observed yield loss.

integrated (Kropff and Spitters, 1992; Kropff *et al.*, 1992b). In two of the three experimental years, drought stress was observed, inducing competition for the resources water and light.

Experimental set-up and results

A series of field experiments was conducted with sugar beet and *C. album* on a sandy loam (2.5 - 5% organic matter, pH 5) in Wageningen, the Netherlands from 1984 to 1986. The soil was at field capacity at 30 vol% water and at wilting point the soil contained 9 vol% of water. The maximum rooting depth was 0.60 m and the groundwater table ranged from 0.60 m early in the growing season to 1.40 m later in the season. The fields were fertilized according to advised amounts of N-P-K for maximal sugar production (Table 8.3). Trials were performed in 1984 with two weed densities (Experiments 1a and b), in 1985 with one density (Experiment 2) and in 1986 with one density and two different dates of weed emergence (Experiments 3a and b, Table 8.3). The crop was seeded with a precision seeder and the weeds were seeded by hand (1985, 1986) or the spontaneously emerging weeds were thinned to the desired densities (1984). All weed removal was done by hand with as little soil disturbance as possible. In all experiments a split plot design was used in four replicates, with the weeds as the main factor and harvest dates as the subfactor.

The date of 50% emergence of the species was estimated from emergence frequency data, based upon daily observations of emergence in permanent plots (2 - 4 m of a row). Measurement of plant-height and observations on crop and weed development were performed at time intervals of about three days.

Above-ground dry weights and leaf area of the crop and the associated weeds were measured at 1 - 2 week intervals throughout the growing season. The samples were divided into various plant organs, dried at 80 °C and weighed. Leaf area of green leaf blades was measured.

At time intervals of about one month, soil samples were taken and dried to determine the moisture content of 10 cm thick soil layers. From these data and experimentally determined pF-curves the volumetric soil moisture content of the soil layers was calculated.

Table 8.3. Details of five field experiments with mixtures of sugar beet (*Beta vulgaris* L.) and *Chenopodium album* L. conducted at Wageningen, the Netherlands in 1984, 1985 and 1986. Source: Kropff *et al.*, 1992b.

	1984		1985	1986	
Experiment	1a	1b	2	3a	3b
Sugar beet cultivar	Regina	Regina	Monohil	Salohil	Salohil
Density sugar beet (pl. m^{-2})	11.11	11.11	11.11	11.11	11.11
Density *C. album* (pl. m^{-2})	5.5	22.2	5.5	9.1	9.7
Sowing date sugar beet	17/4	17/4	24/4	18.4	18/4
Sowing date *C. album*	-	-	14/5	6/5	20/5
Emergence sugar beet	27/4	27/4	9/5	4/5	4/5
Emergence *C. album*	27/4	27/4	19/5	25/5	3/6
Between-row spacing (m)	0.30	0.30	0.30	0.50	0.50
Within-row spacing (m)	0.30	0.30	0.30	0.18	0.18
Gross plot size (m^2)	1.5×3.0	1.5×3.0	1.5×6.0	1.3×6.0	1.30×6.0
Net plot size (m^2)	0.9×1.8	0.9×1.8	0.3×4.8	0.54×4.0	0.54×4.0
No. of replicates	4	4	4	4	4
No. of harvests	7	7	8	10	10
Date final harvest	14/10	24/7	25/9	15/9	15/9
Fertilizer (kg ha^{-1})					
N	160	160	160	170	170
P$_2$O$_5$	40	40	60	38	38
K$_2$O	100	100	200	280	280
Groundwater table (m)	0.7-1.0	0.7-1.0	0.7-0.8	1.2-1.4	1.2-1.4

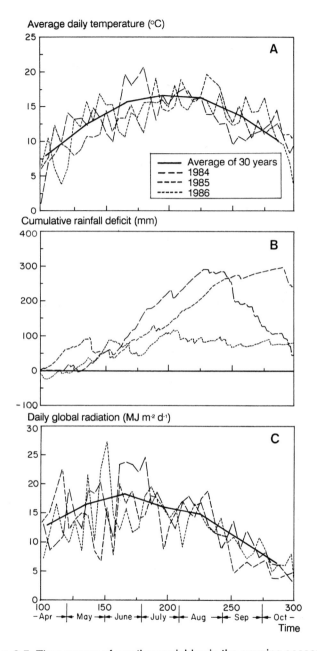

Fig. 8.5. Time course of weather variables in the growing seasons of 1984, 1985 and 1986. (A) Average daily temperature (5 day averages; °C); (B) cumulative rainfall deficit (mm); (C) global daily radiation (5 day averages; MJ m^{-2} d^{-1}). Source: Kropff *et al.*, 1992b.

Experimental details are given in Table 8.3.

Weather conditions differed markedly between the three growing seasons (Fig. 8.5). The most extreme periods were the relatively cold period in May 1984 (days of year 121 - 151) during early growth of the crop and weeds and the relatively warm and bright period in June 1986. The rainfall deficit (Penman evaporation minus rainfall) in 1984 and in 1986 accumulated to about 300 mm, during the growing season. In 1986, however, the plots were irrigated.

Biomass reduction in sugar beet as a result of weed competition varied strongly between the experiments (96% to –6% for total dry matter production; Table 8.4). A significant reduction of crop yield in the weedy plots at final harvest was observed in 1984 (Experiment 1a), 1985 (Experiment 2) and in the treatment with weeds emerging 21 days after the crop (Experiment 3a) in 1986. A period of 30 days weed-free after crop emergence (Experiment 3b) appeared to be long enough to avoid any significant reduction in crop yield.

Total dry matter production of *C. album* was reduced significantly in the mixture plots in all experiments. This reduction was stronger in experiments where sugar beet production was less influenced.

The maximum leaf area index (*LAI*) of the species was strongly reduced in the mixtures (Table 8.4). For sugar beet the reduction in maximum *LAI* ranged from 13 to 89%, whereas the reduction for *C. album* ranged from 22 to 97%.

Maximum height of sugar beet during the growing season was hardly influenced by the weeds in the mixtures; only in 1984 (Experiment 1a) sugar beet height was reduced by 27% (Table 8.4). Maximum height of *C. album* was only influenced in both experiments in 1986 when the weeds emerged 21 - 30 days after the crop. This was obviously due to the strong competitive status of sugar beet with respect to the much later emerged weeds in this experiment. The time course of the available amount of soil moisture was hardly affected by the competition treatments (data not shown). Soil moisture content was slightly more reduced in the mixtures in 1984 compared with the monocultures. The coefficient of variation of the soil moisture content in time of all treatments together was less than 3%.

Simulation analysis of sugar beet - *C. album* competition

To analyse the explanatory power of the model, simulations were performed with the same single set of parameters for all five experiments. The only experiment-specific inputs that were varied between the simulation runs were: daily values of maximum temperature, minimum temperature, global radiation and rain/irrigation, the dates of emergence of crop and weeds and the plant density of the

Table 8.4. Growth data of sugar beet (*Beta vulgaris* L.) and *Chenopodium album* L. in monocultures and mixtures. Dry weights of sugar beet organs and *C. album* are given for final harvest; maximum Leaf Area Index (*LAI*) and maximum height of the species are given for the total growing season. Source: Kropff *et al.*, 1992b.

Experiment	1984					1985			1986			
	mono	mono	1a	1b	mono	mono	2	mono	mono	mono	3a	3b
C. album density (pl. m^{-2})	0	22	5.5	22**	0	11.11	5.5	0	9.1	9.7	9.1	9.7
Sugar beet density (pl. m^{-2})	11.11	0	11.11	11.11	11.11	0	11.11	11.11	0	0	11.11	11.11
Period between crop and weed emerergence (d)	-	-	0	0	-	-	10	-	-	-	21	30
Sugar beet												
Beets (t ha^{-1})	44.1a	-	10.1b	-	61.9a	-	33.1b	53.5ab	-	-	45.4b	56.3b
Beets (t DM ha^{-1})	9.8a*	-	2.2b	0.2	14.5a	-	8.2b	12.9a	-	-	11.2b	13.5a
Total dry matter (t ha^{-1})	14.9a	-	3.2b	0.3	23.1a	-	14.6b	20.4a	-	-	18.9b	21.7
Sugar production (t ha^{-1})	6.9a	-	1.0b	-	9.3a	-	5.1b	9.5a	-	-	8.1b	10.1a
Maximum *LAI*	3.8	-	1.2	0.4	5.7	-	4.2	3.8	-	-	2.9	3.3
Maximum height (m)	0.44	-	0.32	0.25	0.62	-	0.59	0.60	-	-	0.60	0.60
C. album												
Total dry matter (t ha^{-1})	-	10.3b	7.7a	6.7	-	14.3b	4.1a	-	14.4b	13.3b	1.9a	0.30
Maximum *LAI*	-	4.6	2.7	3.6	-	4.7	0.9	-	3.8	3.8	0.3	0.1
Maximum height (m)	-	1.03	0.99	1.08	-	1.66	1.45	-	1.59	1.58	0.92	0.45

*   Means within rows, followed by different letters are significantly different at *P*<0.05, as determined by analysis of variance.

**   This treatment was stopped at July 24.

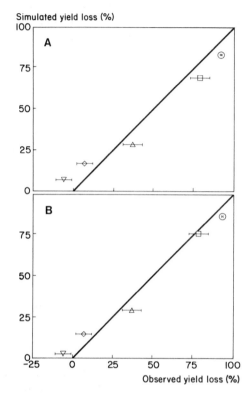

Fig. 8.6. Simulated and observed yield loss (%) for five field experiments with sugar beet and *C. album* in three growing seasons (Experiment 1a (1984): □ ; 1b (1984): o; Experiment 2 (1985): Δ; Experiment 3a (1986): ◊; and 3b (1986): ∇). (A) Simulated with *LAI* as input in the model, (B) *LAI* simulated as well. Source: Kropff *et al.*, 1992b.

crop and the weeds.

The first series of runs was performed with the measured leaf area as input in the model. This enabled the evaluation of the water and carbon balance sections of the model without the confounding effects of feedback between plant growth and leaf area development. The soil moisture content was slightly underestimated by the model. This could be the result of capillary rise. Since this study deals with the analysis of competition it was decided to calibrate the water balance to the measured data of soil moisture content in the monoculture plots of sugar beet, by assuming a capillary rise of 20 mm per growing season. The time course of dry matter production was simulated accurately for both the monocultures and the mixtures in all experiments after this calibration. The simulated yield losses at final harvest with measured leaf area index as input were close to the observed yield losses (Fig. 8.6A).

In the second phase of model evaluation, the routine for simulation of leaf area development (Chapter 4) was activated in the model. Simulation of yield loss was still as accurate as with the measured leaf area as an input (Fig. 8.6B). Total dry weight increment, leaf area development and height development of the species in monocultures and in mixtures were simulated accurately by the model (Fig. 8.7). Sugar beet production in monoculture in 1985 was slightly underestimated by the model and *C. album* production in monoculture was overestimated in 1984 and underestimated in 1986. Model analysis (by simulating additional irrigation) indicated that the production of sugar beet in 1984 was strongly reduced (by 33%) as a result of water stress. In 1986, a reduction in sugar beet production in monoculture of 6% was simulated as a result of water shortage at the end of the growing season whereas in 1985 sugar beet production was not reduced as a result of water stress.

Leaf area development was simulated well for most situations (Fig. 8.7). In the 1984 experiments, the leaf death rate was not simulated accurately, because the strong effect of drought on this process was not included in the model. The impact of competition on height development of *C. album* was quantified satisfactorily using the procedure of a maximum specific stem length at a given height (Chapter 4) (Fig. 8.7). The importance of such a feedback mechanism between plant height and dry matter growth is shown in the 1986 experiment, where height development was strongly reduced in *C. album* as a result of competition effects. Height development was overestimated in the mixture of the 1984 experiment, probably because of the water stress that occurred.

The time course of the soil moisture content was simulated realistically for all experiments (Fig. 8.8). Only at the end of the growing season in 1984 was the soil moisture content overestimated, which did not cause bias in the simulated dry matter production, because the measured soil moisture content was high (no reduction of growth processes).

The model for sugar beet - *C. album* competition was further tested with a completely independent data set in which different weeding treatments were applied (Kropff *et al.*, 1992a). The experimental data used were from de Groot and Groeneveld (1986). Sugar beet var. 'Monohil' was grown at the Droevendaal experimental farm, Wageningen, in 1982 and 1983, at a population of 82000 plants ha^{-1}. A natural weed infestation dominated by *C. album* was allowed

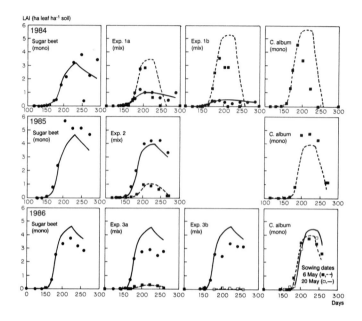

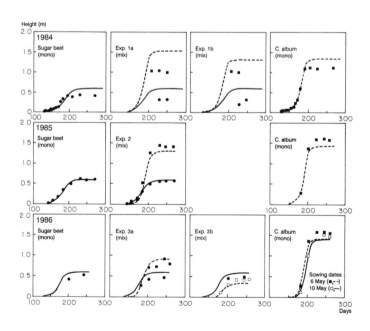

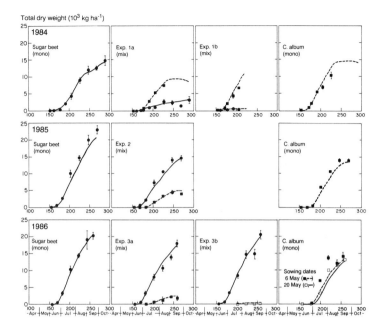

Fig. 8.7. Time course of leaf area development, height development, and dry weight of sugar beet (observed (●) and simulated (solid line)) and *C. album* (observed (■) and simulated (broken line)) in monocultures and mixtures in five experiments performed in three growing seasons, see Table 8.3 for experimental details. Source: LAI and total dry weight, Kropff *et al.*, 1992b.

to develop and remain in the crop for various periods of time, ranging from 20 to 70 days after crop emergence. In other treatments the crop was kept weed-free for different periods of time. A simple module was added to the model that simulated emergence of *C. album* based on the degree-day concept (base temperature of 4 °C and a temperature sum requirement for emergence of 85 °C d).

There was generally good agreement between simulated and observed yields (linear regression of observed versus simulated yield loss: intercept = −0.2, slope = 0.92; $P < 0.001$). The small deviations of model predictions may be due to the existence of other weed species, non-homogeneous emergence of weeds and too low weed densities. In the analysis, a density of 40 plants m^{-2} was assumed, whereas in practice the density may have been higher in some cases.

Soil moisture content (mm)

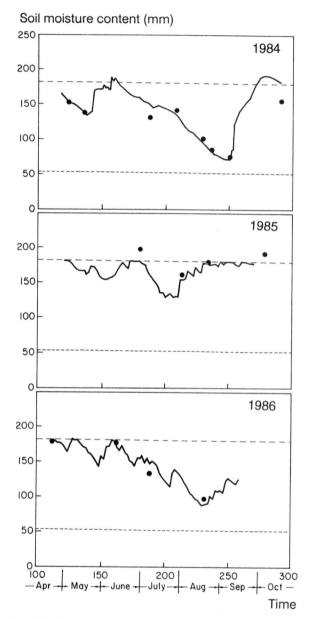

Fig. 8.8. Simulated (solid line) and observed (•) soil moisture content in sugar beet monocultures of three growing seasons. Broken lines indicate soil moisture content of rooted zone at field capacity (pF 2, upper line) and permanent wilting point (pF 4.2, lower line). Source: Kropff *et al.*, 1992b.

Other case studies

Maize and *E. crus-galli*

E. crus-galli has become a difficult to control problem weed in maize crops in the Netherlands. Data on the effect of *E. crus-galli* at different densities on maize yield were published by Kropff *et al.* (1984). 'LG11' maize was grown on a sandy soil at Wageningen, at a population of 110000 plants ha^{-1}, in 1982 and 1983. In 1982, maize emerged on May 15, and in 1983, on June 5. *E. crus-galli* densities were established by thinning natural populations to 0, 100, 200 and 300 plants m^{-2} in 1982, whereas in 1983 naturally established densities were used. In 1982, the weeds emerged five days after the crop and in 1983, the weeds emerged two days before the crop. A detailed simulation analysis was conducted by Kropff *et al.* (1992a). Maize yield under weed-free conditions was accurately simulated during 1982 (observed 13110 kg ha^{-1} ± 1940; simulated 13901 kg ha^{-1}) and 1983 (observed 8440 kg ha^{-1} ± 210; simulated 8459 kg ha^{-1}). The model simulated maize yield losses from *E. crus-galli* competition well in 1982, but greatly underestimated crop yield losses at all weed densities in 1983 (Fig. 8.9). Crop height and leaf area were overestimated by the model in 1983 (data not shown). To determine why the model could not simulate yield loss in the 1983 experiment, an analysis was conducted in which measured height and leaf area were input in the model. Such an analysis indicates if the problem is related to simulation of morphological processes or carbon balance related processes. Simulation results were improved when observed crop and weed heights were input to the model (Fig. 8.9). However, simulations which predicted yield losses similar to those observed in the field were obtained only when the observed leaf area of the crop and weeds at each weed density were introduced to the model. This demonstrates that the model does not (yet) adequately account for the effects of extreme water shortage observed in 1983 on maize morphological development, whereas the impact of drought on growth processes is simulated very well. It was concluded that the combination of drought and *E. crus-galli* competition severely reduced maize stem elongation and leaf area development (Kropff *et al.*, 1992a). In the plots with weeds, the soil moisture content declined to wilting point a few days earlier than in the maize monoculture crop. This could have caused a low soil moisture content exactly at the moment of stem elongation, but direct competition for water may have played an important role as well. This study indicates the complexity of competition in water-limited environments and that process-based models

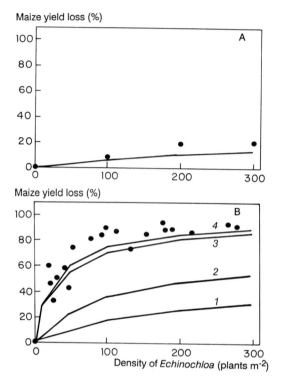

Fig. 8.9. Observed (•) and simulated (lines) yield loss in maize as a result of competition with *E. crus-galli* at different densities in 1982 (A) and 1983 (B). Simulation results for the 1983 experiment were obtained as follows: simulated leaf area and height (line 1), simulated with observed height (line 2) or observed leaf area (line 3) as input, simulated with both observed leaf area and height as input (line 4). Source: Kropff *et al.*, 1992a.

can help to direct research at the process level which is needed for understanding the system. These results demonstrate the importance of extensive model evaluation with data sets from field situations, and using the model as an analytical tool instead of changing species parameters to make the model 'fit' the data.

Tomatoes and *Solanum* spp.

The effect of eastern black nightshade (*S. ptycanthum* Dun.) on transplanted and direct-seeded tomatoes was studied by Weaver *et al.*

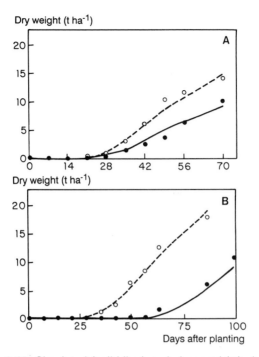

Fig. 8.10. Simulated (solid line) and observed (•) above-ground dry weight of tomatoes and weeds (broken line, o) during the 1981 growing season for (A) transplanted and (B) seeded tomatoes. Source: Weaver *et al.*, 1992.

(1987) in Canada. The model INTERCOM was parameterized for these species and evaluated using experimental data (Kropff *et al.*, 1992a; Weaver *et al.*, 1992). Growth of the species in direct-seeded and transplanted tomatoes was simulated accurately (Fig. 8.10; Weaver *et al.*, 1992). Like in rice, the impact of weeds on direct-seeded tomatoes is much greater than on transplanted tomatoes because of the difference in starting position.

Wheat and a mixture of weeds

The model was also used by Lotz *et al.* (1990) to analyse experimental data on the effect of omitting herbicide applications in winter wheat (*Triticum aestivum* L.) in the Netherlands. Results confirmed the observations in various experiments that weeds can only reduce wheat yield when they emerge shortly after crop emergence in autumn and when they have strong height development. Weeds emerging in spring could not reduce wheat yield significantly even if they were allowed to grow as high as the crop, which is a common observation in highly productive wheat systems in the Netherlands.

Eco-physiological analysis of crop-weed competition for light, water and nitrogen in field situations

In the previous sections it has been shown that an eco-physiological model like the model INTERCOM can predict competition effects for a wide range of situations. However, it was also shown that the model is a simplification of the real system, reflecting our current state of knowledge. That makes the critical test of the model with experimental data crucial. Because the model is based on physiological processes and their response to the environment, the model can be used to analyse the experimental results in more detail. A series of questions can now be raised like: What is the relative contribution of competition for the resources light, water and nitrogen? Would that be different if the densities or dates of weed emergence were different than in the experiments? A second series of questions that are important for understanding the background of competition is related to the relative contribution of factors like weather, weed density and the period between crop and weed emergence, and to differences in yield loss as observed in the experiments.

The first question that has to be discussed is related to the contribution of competition for nitrogen. The model version of INTERCOM we used in the sugar beet - *C. album* and the rice - *E. crus-galli* studies did not contain a nitrogen soil balance, uptake and distribution component. However, we observed in both experiments no differences in the leaf N concentration throughout the growing season and leaf N was definitely not at a low critical level (Fig. 8.11). This indicates that in these systems competition for capture of the resources light and water determined the growth rate of the plants and by that the demand for N which could be supplied by the soil. In those situations, N levels may be the same in monocultures and mixtures, but because it is at the minimum level.

To analyse the contribution of competition for water to the observed variation in yield loss, simulation runs were performed for all sugar beet experiments with optimal water supply by introducing a simulated irrigation of 4 mm d^{-1}, whereas the other variable inputs remained unchanged. Although the simulated yield in monoculture increased for 1984 and 1986, by 33% and 6% respectively, simulated

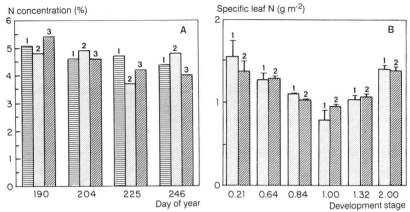

Fig. 8.11. (A) Nitrogen concentration during the growing season in leaves of sugar beet growing in monoculture (1), a mixture with 5.5 and 22 *C. album* plants m^{-2} (2 and 3, resp.); 1984 experiments (see text). (B) Specific leaf N (g m^{-2}) during the growing season in leaves of rice growing (1) in monoculture and (2) in mixture with *E. crus-galli*; 1992 dry season experiment, IRRI, Los Banos, Philippines.

relative yield losses were almost equal to simulated relative yield losses for the drought stressed real situations.

The effect of water shortage on competitive relationships was further analysed by the model by using average weather data for Wageningen and assuming no rainfall, with a soil profile at field capacity at the beginning of the growing season. The effect of three irrigation treatments was evaluated with the model at three maximum weed height and two dates of weed emergence. Maximum weed height and the period between crop and weed emergence strongly influenced the competitive strength of the weeds at all irrigation levels (Table 8.5). When the weeds grew twice as high as the crop (as is mostly the case in sugar beet - *C. album* mixtures), water shortage had little effect on yield loss. A slight increase in yield loss with increasing irrigation level was simulated when the weeds emerged simultaneously with the crop and a slight reduction in yield loss was simulated when the weeds emerged 10 days after the crop. However, when the weeds were unable to overtop the crop, soil moisture shortage strongly increased the competitive effect of the weeds. This effect was more marked when the weeds emerged earlier.

In the situation where the maximum height of the weeds was half that of the crop, the competitive strength of the weeds with respect to light was small. However, the weeds remove extra water from the soil compared to the monoculture, especially early in the growing season when the canopy was still open. Water shortage will occur earlier in the season when weeds are present, resulting in considerable yield loss as a result of water competition. Because the weeds were taller than the crop, water shortage hardly influenced the outcome of competition in the specific experiments.

The accuracy of the simulation results mentioned in the previous sections allow model analysis of the backgrounds of the differences in yield loss between the experiments by varying dates of weed emergence, weed density and weather data. This was done for the detailed study with sugar beet and *C. album*.

The differences in yield loss between the treatments in Experiment 1 and Experiment 3 were caused by only a single factor, density and date of weed emergence, respectively. The higher weed density in Experiment 1b (1984) with respect to Experiment 1a caused a severe increase in yield reduction of the sugar beet. The later weed emergence in Experiment 3b with respect to Experiment 3a (1986) resulted in a smaller yield reduction. As the model performed well for these situations, it can be concluded that the effects of weed density and the period between crop and weed emergence on crop yield loss can be simulated precisely with the model. Differences in the effect of *C. album* on sugar beet yield between the years, however, were due to a combination of factors: weather, weed density, and dates of crop and

Table 8.5. Simulated yield losses of sugar beet in competition with *C. album* for three irrigation levels. Three relative maximum weed heights of the weeds with respect to the crop and two relative times of emergence of the weeds were assumed. Irrigation levels: 0, 50 and 100 days after emergence at 7 mm d^{-1}. Weed density was 11 plants m^{-2}, growth period 150 days, soil moisture content initialized at field capacity, average weather data Wageningen, without rainfall. Weed-free yield of sugar beets for non-irrigated, 50 and 100 days irrigation was 12.4, 14.0, 24.1 t dry matter ha^{-1}, respectively.

	Period between crop and weed emergence (d)					
	0			10		
Irrigation period (d)	0	50	100	0	50	100
Relative maximum height of weeds						
2	80	79	77	52	53	54
1	74	70	62	20	19	16
0.5	24	15	8	7	4	2

weed emergence. The contributions of weed density, the period between crop and weed emergence, radiation and temperature were analysed at potential conditions with an optimal water supply.

Because of the excellent performance of the model, the model was used for further interpretation: the roles of weed density, the period between crop and weed emergence, temperature and radiation in the experiments were analysed by comparing simulated crop yield using observed weed densities, periods between crop and weed emergence and weather, with simulated yields based on the average weed density, period between crop and weed emergence and weather. Yield loss based on average inputs was simulated to be 38% (Table 8.6). Step by step the average values were replaced with the experimental specific values in two sequences. Table 8.6 shows that differences in weed density between the experiments only explained 12% of the observed variation, whereas differences in the dates of weed emergence accounted for 96% of the variation. This was the result of very small differences in weed density in the experiments. Introduction of the observed radiation and temperature improved the simulations only slightly. This analysis is illustrated in Fig. 8.12.

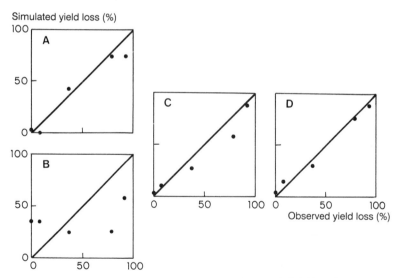

Fig. 8.12. Relation between simulated and observed yield loss in sugar beet - *C. album* mixtures based on average inputs across experiments, but with (A) actual dates of emergence, (B) actual densities, (C) actual dates of emergence and densities and (D) actual dates of emergence, densities and weather.

The influence of weather variables on yield loss and its interaction with the period between crop and weed emergence were studied with the model in a third series of simulations (Table 8.7). When the weeds emerge simultaneously with the crop, yield loss is hardly influenced by radiation and temperature. However, when the weeds emerge after the crop, yield losses were simulated to be much higher when simulated with 1984 weather data. Analysis of model output showed a much slower leaf area development with these weather data, which resulted in less advantage of the crop with respect to the weeds in comparison with the other weather data. This was due to the relatively low temperature in the period after emergence in 1984 (Fig. 8.5). This leads to the conclusion that the period between crop and weed emergence in simple models should not be expressed in days but in a developmental measure such as degree days (when the temperature response of leaf area development is linear).

Table 8.6. Interpretation of the backgrounds of differences in yield loss between five field experiments with sugar beet and *C. album* with the model. Observed and simulated yield loss (% of total dry weight of sugar beet in monoculture) and the % of the variation accounted for with the model (r^2) are given. After a standard run with average experimental specific inputs without water shortage, the average inputs for dates of weed emergence, weed density and other experiment-specific variables (weather variables) were changed into measured data at different sequences (A, B). Source: Kropff *et al.*, 1992b.

	1984		1985	1986		r^2
Experiment	1a	1b	2	3a	3b	
Observed yield loss (%)	79	93	37	7	-6	
Simulated yield loss (%)						
- Standard run with av. data	38	38	38	38	38	
- A. +Dates of emergence	74	74	42	0	1	0.96
+Weed density	57	87	27	10	1	0.96
+Weather	77	88	30	15	3	0.98
- B. +Weed density	25	58	25	35	36	0.12
+Dates of emergence	57	87	27	10	1	0.96
+Weather	77	88	30	15	3	0.98

Table 8.7. Analysis of the effect of radiation and temperature and relative time of weed emergence on sugar beet yield loss (% of total dry matter) as a result of competition with *C. album*. Weather data were from Wageningen, 1984, 1985, 1986 and 30 years average, water was available in ample supply, the growth period was 150 days, *C. album* density was 5.5 plants m^{-2}. Relative times of weed emergence correspond with the values of the five experiments: 0 (2x), 12, 21, 30 days, respectively. Source: Kropff *et al.*, 1992b.

	Period between crop and weed emergence (d)			
	0	12	21	30
Weather data				
1984	67	53	34	12
1985	63	29	6	0
1986	61	34	10	2
1951-1980	63	39	15	2

Discussion and conclusions

The results presented in this chapter demonstrate the potential of eco-physiological process based models to improve our understanding of competition effects in field situations. Such detailed models must be seen as a research tool, that can be used in the analysis of factors of importance to competition. They reflect the current state of knowledge at the process level and serve as vehicles to integrate that knowledge and link it to the field level. It depends on the objectives of users/developers if it is worthwhile to conduct specific research to fill up gaps in knowledge, like the effect of severe drought on maize leaf area development. If such drought situations are of major importance to farmers, it might be useful to obtain more insight so that the model can predict yield loss in these situations as well.

The model INTERCOM has been constructed from procedures used in general crop growth models, and competition was introduced by distributing the resources light and water over the species. Basically, the physiological processes are the same irrespective of whether a species is grown in monoculture or in competition with other species. However, in the latter situation, the attributes determining resource interception become more important because they regulate the distribution of the limiting resources between the competing species. For a monoculture crop in the linear growth phase, light capture hardly increases with *LAI* above 4 m^2 m^{-2}, and plant height does not affect light capture at all (analysed by the model INTERCOM, data not shown). In a mixture, however, small differences in plant height or leaf area development may cause dramatic changes in the competitive relationships. The attributes that determine the capacity of a species to capture resources, like starting position, plant height, leaf area dynamics and root morphology, need, therefore, special attention in a competition model. Several of these aspects might be improved in the model, based on specific research. For drought and nutrient limited situations it would be useful to obtain more insight in root-related processes. Height development and its plasticity in competition situations can also be identified as an area of great importance in almost all competition situations. Further refinement, however, will involve intensive physiological research.

Interaction between different species is described in the model based on the assumption that neighbouring plants reduce each other's growth only by modifying the environment, by changing the availability of light and water. The results presented in this chapter indicate that this assumption does not cause major under- or overestimation of competition effects, because the behaviour of species in mixtures could be understood from their behaviour when growing alone in monoculture.

The model INTERCOM assumes a horizontally homogeneous leaf area distribution, which may cause deviations when row spacing of the crop is large. This can be overcome by simulating growth in competition on a per plant basis. For each plant its light absorption can be calculated based on light absorption of the plant and its neighbours. A model for this purpose was developed by R. Stokkers, M.J. Kropff, J. Goudriaan and J.A. den Dulk (Department of Theoretical Production Ecology, Wageningen Agricultural University). Detailed light distribution models were developed by van Kraalingen (1989) and Gijzen and Goudriaan (1989) and could be used for that purpose. However, these models require much computation time and detailed information on the 3-dimensional structure of the system. A realistic possibility to account for low density effects or heterogeneously distributed weed populations is to distinguish smaller areas within fields with different weed densities and simulate yield loss for these areas separately. Afterwards yield loss can be averaged, weighted by the proportional area at each density. The approach is similar to the approach proposed by Wilkerson *et al.* (1990). This aspect is discussed in more detail in Chapter 3.

This approach was evaluated by running the model INTERCOM (sugar beet - *C. album*) for low weed densities and a situation where the weeds are clustered but with the same number of plants per hectare (1000 plants ha^{-1}). This analysis showed that this effect can be significant if the weeds are strongly clustered: a homogeneous distribution resulted in a yield loss of 1%, but if the weeds were only growing on 5% of the area, the yield loss was 0.8%

The range of environmental conditions encountered by a species when grown in mixtures is much broader than when grown in monoculture. For instance, a short weed species experiences a low light intensity and low red/far red ratio when grown with a tall crop. Such a micro-environment usually increases its stem elongation rate, specific leaf area, and shoot/root ratio and reduces its light-saturated photosynthesis and maintenance respiration. Plants and especially weeds have a large plasticity due to this type of response. The effect of competition on stem elongation of *C. album* was incorporated in the model INTERCOM by a very simple empirical approach using the maximum Specific Stem Length (SSL) as a function of height of the plant. This can be seen as a very simple way to account for the phenotypic plasticity of this species with respect to height development.

Eco-physiological models for interplant competition that are similar to the model discussed here were developed by Wilkerson *et al.* (1990) and Graf *et al.* (1990a, b). Wilkerson *et al.* (1990) developed their model on the basis of a soybean growth model (Wilkerson *et al.*, 1983), that simulates crop growth in a more simplified way than the model INTERCOM. Competition is simulated by defining an 'area of influence' for each weed, that is the area where the weeds compete with the crop to account for the horizontally heterogeneous distribution of the weeds. The field average of light interception is calculated based on the proportions of the field that are inside and outside weed areas of influence. However, the area of influence is not only dependent on the diameter of the plant, but also on plant height, a factor which is not accounted for. Wilkerson *et al.* (1990) derived a light distribution factor by fitting the model to growth data of the species in mixtures. So, the effect of height, one of the key factors, is included in an empirical way in their model. The model accurately simulated the time course of dry matter for soybean (*Glycine max* (L.) Merr.) and common cocklebur (*Xanthium strumarium* L.) for two seasons at the same site in which competition effects were very similar. An evaluation of the model for contrasting competition situations would be helpful.

The model of Graf *et al.* (1990b) for nitrogen and light competition in rice is based on a general crop growth model developed by Graf *et al.* (1990a). Many aspects of their model are similar to the approaches used in the model INTERCOM. They divided the weed flora into six groups based on differences in leaf shape, growth form, height and phenology. Graf *et al.* (1990a) also found a close correspondence between simulated and observed dry matter production of crop and weeds. The effect of different weeding treatments was simulated accurately. Graf and Hill (1992) successfully evaluated the same model using a detailed data set for rice - *E. crus-galli* competition. A comparison of Graf's model with the model INTERCOM using the same data set would be useful.

The model INTERCOM is a comprehensive one, directed towards understanding the basic principles governing crop-weed interactions. It is, therefore, intended as a research tool. Such a model also facilitates the study of plant attributes determining the competitive ability of weeds, and the evaluation of weed control strategies. An example is the quantification of non-lethal weed control measures such as bio-herbicides in terms of yield loss, and low dosages of herbicides if their effect on physiological and morphological characteristics is known.

Decision-support systems for weed control require competition models generating accurate and reliable predictions of the crop yield reduction to be expected, whereas insight into the mechanisms behind the yield reduction is of lesser interest. Therefore, detailed models, such as the model INTERCOM, may be less suitable for that purpose. The detailed models are, however, very well suited to derive simple, predictive models. These applications are discussed in Chapters 9 and 10.

Chapter Nine

The Impact of Environmental and Genetic Factors

M.J. Kropff, N.C. van Keulen, H.H. van Laar and B.J. Schnieders

Introduction

Crop-weed competition in real systems is a complex phenomenon because many factors determine the outcome of the process, even in simple situations where plants only compete for light capture. Factors that are well recognized as being important are: plant density, the period between crop and weed emergence, weather conditions and plant traits like early growth rate, photosynthetic rate, morphological traits like height and leaf angle arrangement, seed size, water and or nutrient use efficiency. Pheno-typic plasticity of the species makes the system even more complex. A qualitative indication that any of these factors can be important in specific situations is easy to obtain from experimental evidence in the vast literature on competition. However, a quantitative indication of the competitive ability of a species in relation to the specific combination of the environmental factors and the traits that a plant exhibits is difficult to obtain, as environmental factors are always different between situations. That often results in complex discussions trying to explain why a factor was important in one situation but not in another. It is also impossible to relate differences in competitive ability between varieties to single traits in an experimental way, because varieties always differ for a combination of traits. This could only be achieved by using isogenic lines.

However, once a model with enough eco-physiological detail, like the model INTERCOM, is operational and validated, it can be used as a tool for such quantitative analysis and serve as a vehicle to summarize the vast knowledge we have on crop-weed competition mechanisms. Isogenic lines of a species can easily be generated in the model for analysis and quantification of the effect of individual traits on the competitive ability of the species. That can be done by changing one parameter value at a time for the specific trait under study like maximum height. The model can also facilitate the analysis of such effects in a range of environments, emergence dates, weather conditions and densities, as the importance of a specific factor is related to the specific competition situation.

In this chapter, we present the results of some of these applications of the simulation model INTERCOM like the quantification of the effect of plant density, the period between crop and weed emergence, weather and plant traits on competition. However, the ultimate application of the model would be the use of it by scientists who have specific hypotheses related to competition who would like to pre-test that hypothesis, which may help to improve the experiment.

Plant density and the period between crop and weed emergence

Both weed density and date of emergence of the species largely determine crop yield loss by weed competition. The model INTERCOM was used to study the combined effect of both factors on yield loss, for the well-tested competition situation with sugar beet and *C. album*. As shown in Chapter 8, *C. album* can grow twice as high as the sugar beet crop and can cause very high yield losses at very low weed densities if the period between crop and weed emergence is short. Experimental determination of the complete response surface of yield loss as a function of density and period between crop and weed emergence is extremely laborious. Because the model INTERCOM was evaluated for extreme situations, it should be able to simulate a realistic response surface. Yield loss was simulated using 1985 weather data (see Chapter 8) for a wide range of weed densities (0.5 to 64 plants m^{-2}) and at all densities for different periods between crop and weed emergence (–5 to 40 days). The results are presented in Fig. 9.1C. The response surface shows a strong interaction between the effect of the two factors. Weeds that emerge simultaneously with the crop or shortly after the crop cause severe yield losses at very low densities (Fig. 9.1C). However, when the weeds emerge 25 days or more after the crop, yield losses become very small even at high densities. These results show the importance of the time of emergence of the competing species.

In a subsequent analysis, the effect of maximum weed height on the response surface was analysed. The effect of weeds with all the traits of *C. album*, but differing only in the maximum attainable height, on sugar beet yield loss was simulated. Maximum height of the weeds strongly affected the response surface (Fig. 9.1). If the maximum weed height was reduced to 60 cm (which equals the maximum height of the crop), only early emerging weeds could cause

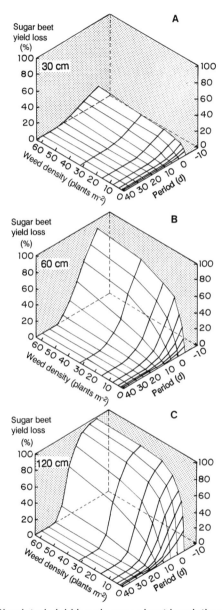

Fig. 9.1. Simulated yield loss in sugar beet in relation to weed density and the period between crop and weed emergence for *C. album*. The weed species has all morphological and physiological characteristics of *C. album*, except maximum plant height (A: 30 cm; B; 60 cm; C; 120 cm).

severe yield loss. If the maximum height was reduced to 30 cm, only small yield losses were observed by early emerging weeds.

The results also shows the pitfalls in experimental analyses of the importance of the period between crop and weed emergence, if only a few treatments are applied. In the situation discussed here, yield loss is extremely sensitive to small changes in the period between crop and weed emergence between 10 and 20 days after crop emergence for the normal *C. album* in which yield loss increases from 20% to 90% at high densities. For the other simulated isolines of *C. album* which differ in maximum weed height, the sensitivity is smaller and the timing is different (Fig. 9.1). This has strong implications for experimental analyses in which a reasonable effect is needed to obtain effects that can be detected by statistical analysis.

These data, generated by the simulation model INTERCOM, were subsequently used to evaluate the performance of the regression equation for the effect of both weed density and the period between crop and weed emergence on yield loss that was proposed by Cousens *et al.* (1987) (Eqn 2.9). This equation has huge data needs, because both factors are involved. Such data sets are difficult to obtain experimentally. This regression equation accurately described simulated yield losses at different densities and dates of weed emergence (Fig. 9.2). However, the regression equation was unable to describe the effect of very early emerging weeds (Fig. 9.2). The equation assumes a constant maximum yield loss, which does not depend on density, as the period between crop and weed emergence becomes large and negative. Such situations will seldom occur in normal tillage situations. However, in no tillage systems and in semi-natural systems, this discrepancy may be of concern. The result is a series of parallel curves. Largest differences were found in the parameter x that indicates yield loss per weed at low densities when the crop and weed emerge on the same day (Table 9.1).

Table 9.1. Estimated parameter values (see Eqn 2.9) and standard errors using the regression models of Cousens *et al.* (1987) to fit simulated data on the effect of *C. album* competition on sugar beet yield losses with three different maximum weed heights. Source: Kropff *et al.*, 1992a.

Weed height (cm)	df	Parameter values						r^2
		z		x		y		
30	51	98.9 $\pm$	5.0	0.44 $\pm$	0.02	0.136 $\pm$	0.003	0.98
60	51	98.2 $\pm$	7.7	5.55 $\pm$	0.35	0.132 $\pm$	0.005	0.97
120	51	93.1 $\pm$	15.2	53.88 $\pm$	8.41	0.167 $\pm$	0.008	0.96

Sugar beet yield loss (%)

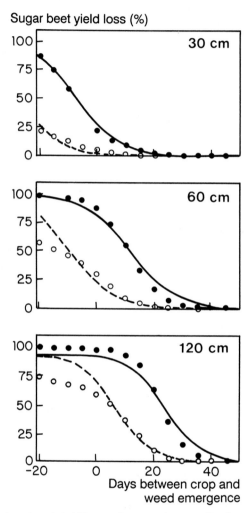

Fig. 9.2. Simulated yield losses in sugar beet at *C. album* densities of 5.5 plants m^{-2} (o) and 88.0 plants m^{-2} (•) for a wide range of periods between crop and weed emergence. Maximum height of the weeds was 30, 60 or 120 cm, maximum height of the crop was 60 cm. Lines present the result of fitting the simulated data, at a wide range of densities between 5.5 and 88.0 plants m^{-2} with the regression model of Cousens *et al.* (1987). Source: Kropff *et al.*, 1992a.

Climatic factors

The use of the model INTERCOM to determine the impact of climatic factors on crop-weed competition has already been discussed in Chapter 8 where the backgrounds of the different competition effects were analysed.

The effect of water shortage on crop-weed competition at different irrigation periods and dates of weed emergence has already been demonstrated for three weeds, only differing in height (Table 8.5). It was shown that when the weeds are taller than the crop, yield loss hardly differed among irrigation regimes because competition for light was the dominant process determining yield loss. It is important to note here that simulations started with a saturated soil profile which enabled the development of a closed canopy. However, weed species that were unable to overtop the crop and, thus, were unable to cause severe yield reduction by light competition, reduced simulated crop production more strongly in water limited conditions compared with well watered conditions. This was the result of increased water losses in the mixture compared to the monoculture, which caused growth reductions in sugar beet as a result of water shortage later in the season.

Tilman's theory (1988) states that plants are more competitive if they have a higher tolerance to low resource levels. That would mean that a higher soil depletion factor for water (Eqn 5.15) would lead to an increased competitive ability of the sugar beet. This was evaluated by the model INTERCOM for the year 1984. Yield loss in sugar beet was reduced from 52% to 50% by 5.5 plants m^{-2} if the sugar beet had the characteristics for the soil depletion factor of a C_4 species and the weeds grew as high as the crop. Weeds that grew only half as tall as the crop caused a yield loss of 15%, but with a higher efficiency of the crop yield loss was 13%. This indicates that the effect of the soil depletion factor in such situations is not large probably because competition for the resource water was not that important. Another reason for the small impact will be the timing of the drought period which was late in the season (Fig. 8.8). Other effects like a difference in rooted depth (niche differentiation) or differences in effective root length can only be studied with a more complex water balance model (Chapter 5). These results confirm Tilman's theory. However, his theory will be more relevant for situations where resource levels are very low (Grace, 1990).

The effect of temperature on competition effects was also analysed in Chapter 8 (Table 8.7). It was concluded that temperature plays a very important role in the period between crop and weed emergence, because the temperature determines the leaf area development of the first emerged species. So, it determines the starting position of the first emerging species at the time the other species emerge. The impact of temperature was further studied by the model INTERCOM by changing the daily temperature by + or − 2 °C for the sugar beet - *C. album* and rice - *E. crus-galli* situations. For sugar beet and *C. album*, the effect of temperature was clearly dependent on the length of the period between crop and weed emergence (Fig. 9.3A). These results indicate that an increase in temperature as a result of global change will lead to reduced yield losses by weeds, especially by late emerging weeds. At higher temperatures, the advantage of the crop when the weeds emerge increases. This will generally hold for C_3 crops and weeds in temperate climates. If a C_3 crop species competes with a C_4 weed species (like *E. crus-galli*), an increase in temperature will lead to higher yield loss in early emerging weeds, as photosynthesis in C_4 species benefits from higher temperatures.

A change in temperature on competition between rice and *E. crus-galli* hardly affected yield losses by *E. crus-galli*. Yield losses slightly increased at a higher temperature, because *E. crus-galli* is a C_4 species, which has higher optimum temperatures for several processes (Chapter 4) (Fig. 9.3B).

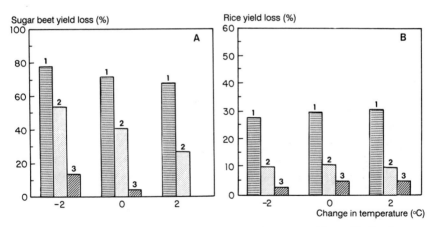

Fig. 9.3. Simulated yield loss in (A) sugar beet by competition with *C. album* (5.5 plants m^{-2}, 1985, Wageningen, the Netherlands) and (B) rice (700 plants m^{-2}, direct-seeded) by competition with *E. crus-galli* (100 plants m^{-2}, Los Banos, Philippines, 1992) at different periods between crop and weed emergence at normal temperatures, and a temperature change of −2 °C and +2 °C for the whole growing season. Periods between crop and weed emergence: 0 (1), 10 (2) and 20 days (3).

The effect of changes in CO_2 concentration can also be analysed by the model. The effect of CO_2 on the initial light use efficiency and the maximum rate of leaf photosynthesis can be introduced in the model according to functions like those indicated by Jansen (1990). It is well known that C_4 species show much less photosynthetic response to CO_2 enrichment than C_3 species, as a result of differences in the biochemical pathways. So, increased CO_2 may be beneficial to C_3 species, but will not be beneficial for C_4 species. Differences in morphological responses to CO_2 between species like changes in the Leaf Area Ratio have been observed (Bazzaz *et al.*, 1989). Such responses will strongly affect the competitive ability of species.

Morphological and physiological species characteristics

Relationships between morphological and physiological traits of species and their competitive strength have been widely studied (cf. Black *et al.*, 1969; Pearcy *et al.*, 1981; Rooney, 1991; Légère and Schreiber, 1989; Pons, 1985). Pearcy *et al.* (1981) found a close relationship between competitive ability and photosynthetic responses of *C. album*, a C_3 species, and *Amaranthus retroflexus*, a C_4 species. At all conditions studied, high competitive advantage was coupled to higher photosynthetic capacity. Rooney (1991) concluded that the growth form differences between *Avena fatua* lines were mainly responsible for differences in competitive ability versus wheat. The complexity of relationships between morphological and physiological characteristics and competitive ability of plants have been recognized by many researchers (cf. Pons, 1985). However, qualitative or quantitative studies on the impact of physiological differences on competitive ability of weeds are rare. Also here, an eco-physiological simulation model for competition between plant species for light and water like INTERCOM can be of help. The model links morphological and physiological characteristics of species to growth in dependence on the environment in a mechanistic way. This model can, thus, be used to analyse the impact of different morphological and physiological traits on the competitive ability of a species.

A wide range of competition situations can be generated by the model. However, because the model was most intensively tested for sugar beet - *C. album* competition, an analysis of the importance of such traits was conducted for that competition situation, including the different isolines of *C. album* with a different maximum height. Daily weather input variables were the 30 year average data for maximum temperature, minimum temperature and global radiation

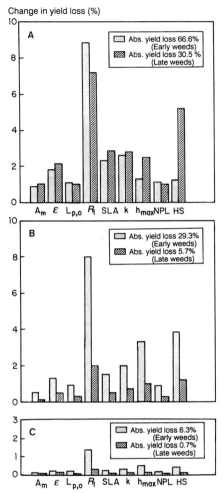

Fig. 9.4. Effect on changing parameter values of the weeds by 5% on simulated yield loss in sugar beet in competition with *C. album* (A), and weed species with all morphological and physiological characteristics of *C. album*, except maximum plant height which was equal to the height of sugar beet (60 cm, B) or 30 cm (C). The change in yield loss is indicated as the yield loss after the change minus the standard yield loss.

for Wageningen, the Netherlands. Water was assumed to be available in ample supply by introducing irrigation in the model. The sensitivity of simulated yield loss in sugar beet to several morphological and physiological species characteristics was studied by

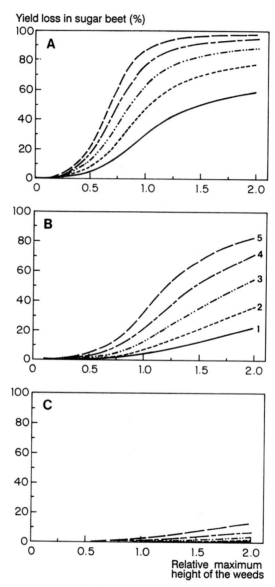

Fig. 9.5. Simulated yield loss in sugar beet by competition with *C. album* L. at different densities in relation to relative maximum height of the weed species when the weeds emerge simultaneously with the crop (A), and when the weeds emerge 15 days (B) or 30 days (C) after the crop. Lines 1-5 represent 5.5, 11, 22, 44 and 88 plants m^{-2}, respectively.

increasing the values of the listed weed characteristics by 5% (Fig. 9.4). In the analysis, the sensitivity was studied at different periods between crop and weed emergence and weed densities, as different factors may be important in different competition situations. The sensitivity follows from the change in yield loss as presented in Fig. 9.4. The relative sensitivity expressed as the ratio of % change in yield loss and % change in the value of the specific parameter was reported by Kropff *et al.* (1992b).

Simulated yield loss ranged from 0.7 to 67% for the two situations in which the weeds had all species characteristics of *C. album*, which grew twice as high as the sugar beet crop (for comparison with actual yield loss see Chapter 8). The effect of weeds with the same morphological and physiological characteristics of *C. album* except for maximum height, was considerably smaller: 6 - 29% yield loss when the maximum height of the weed was equal to the value of the crop and 0.7 - 6% yield loss when the maximum height of the weed was half the value of the crop.

Simulated yield loss was most sensitive to the morphological parameter R_l, which determines the rate of leaf area development in early growth phases in dependence on temperature, and the parameter s which determines the earliness of height development in all competition situations. Simulated yield loss was less sensitive to physiological species characteristics (parameters related to photosynthesis) than to morphological parameters (parameters determining leaf area development, plant height and efficiency of light absorption). These results indicate the importance of accurate measurements of these parameters for prediction of the effects. On the other hand a strong sensitivity indicates large opportunities for breeders to develop varieties with a higher competitive ability. These results demonstrate the non-linear and strongly interrelated relationships between plant traits and their impact on competitive ability.

The variability in sensitivity of simulated yield loss as a function of the relative maximum height of the weed species and the period between crop and weed emergence is illustrated in Fig. 9.5. The simulated results demonstrate that the sensitivity of yield loss (slope of the curves) to the relative maximum height of the weed and its timing depends upon the relative maximum height of the weed itself and also on weed density and the period between crop and weed emergence. These complex interrelationships indicate that experimental evaluation of the importance of such factors is extremely difficult. An eco-physiological model can be helpful in integrating the effects of these traits.

Chapter Ten

Practical Applications

M.J. Kropff, L.A.P. Lotz and S.E. Weaver

Introduction

Competition between plants is a complex process that determines the performance of (semi-) natural and agricultural systems. In agricultural research, it is of great importance that the knowledge of the mechanisms of competition is used to improve crop management practices. Competition studies in weed science have been fairly empirical for a long time. The model INTERCOM has been developed and used to improve insight into the mechanisms of competition between arable crops and weeds. However, the approaches discussed in this book can be used to study interplant relationships in other systems as well. The model could, for example, be useful in obtaining insight into how the available resources could be used more efficiently by changing relative times of emergence, densities and cultivars in intercropping systems. That would require the parameterization of the model for the species that are intercropped. Such models are being developed for intercropping systems by Caldwell and Hansen (1990). In natural or semi-natural ecosystems interrelationships are often more complex than in agricultural systems because of variation in time and space, making it very difficult to develop, parameterize and validate detailed eco-physiological models. Besides that, ecologists are generally especially interested in long-term dynamics of perennial systems. For such situations more simple approaches are more appropriate to study succession or long-term changes in species composition in those systems (e.g. Berendse *et al.*, 1987; Berendse *et al.*, 1989; Tilman, 1988).

Although insight into the process of competition obtained by a detailed model like INTERCOM may be important, the effort to build such models in agricultural research projects becomes useful only if the model can be used to improve the management of agricultural systems.

149

This chapter describes several applications of the eco-physiological model INTERCOM. The pre-testing of new, simple predictive approaches for the impact of weeds on crops is discussed first, followed by the analysis of the impact of non-lethal doses of (bio) herbicides on competitive ability, the quantification of the effect of spatial heterogeneity in weed distribution and the prediction of yield loss in future crops. In a separate section, the widely studied concept of the critical period for weed control is analysed by the model, followed by a discussion on the perspectives for breeders. Finally, a framework for improved weed management systems is presented and the role of systems approaches in weed research is discussed.

Predicting yield loss

Development of simple predictive tools

Decision-support systems for weed control require competition models that generate accurate and reliable predictions of crop yield reduction, whereas insight into the mechanisms is of lesser interest. From a farmer's viewpoint, it may not matter too much that a model is based on mistaken assumptions as long as the model predicts well (Thornley and Johnson, 1990). However, regression models that are biologically meaningful and that contain parameters that have a biological meaning are to be preferred of course (Cousens *et al.*, 1987).

Detailed eco-physiological models, such as the model INTERCOM, are less suitable for prediction purposes. These models require too many inputs and parameter estimates. For practical purposes, simple relationships or rules of thumb are needed to predict yield loss based on simple to conduct observations. The detailed models are, however, very well suited to design new concepts for simple, predictive approaches and to test them before laborious experimentation starts.

Such an alternative predictive approach (to predict yield loss from relative leaf area) was suggested by Spitters and Aerts (1983) and Kropff (1988a). It was shown by simulation studies that a close relationship exists between yield loss and relative leaf area of the weeds, determined shortly after crop emergence for sugar beet and *C. album*, a combination for which the model was thoroughly validated (Chapter 8). In these simulations, both density of the weeds and the period between crop and weed emergence were varied over a wide range. At all weed densities (5.5, 11, 22, 44, and 88 plants m^{-2}) the period between crop and weed emergence was varied over a wide range (–5, 0, 5, 10, 15, 20, 25, 30, 35, 40 days). This series of simulations was also conducted for two theoretical weed species, with all characteristics of

Yield loss (%)

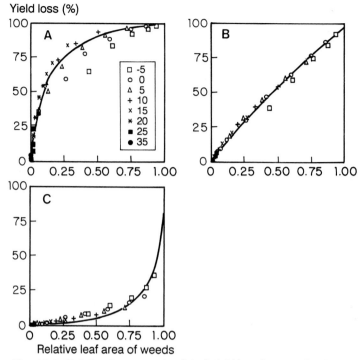

Relative leaf area of weeds

Fig. 10.1. Relation between simulated yield loss in sugar beet and relative leaf area of *C. album* determined 30 days after sugar beet emergence. *C. album* densities were 5.5, 11, 22, 44 and 88 plants m^{-2} at a wide range of periods between crop and weed emergence (see legend in (A)) for weeds with a maximum height of (A) 120 cm, (B) 60 cm and (C) 30 cm. Maximum height of the crop was 60 cm. Lines give the result of fitting the simulated data with a regression model (Eqn 2.18). Source: Kropff and Spitters, 1991.

C. album, but differing only in maximum height. The first theoretical species had a maximum height of 60 cm (like the sugar beet crop) and the second one had a maximum height of 30 cm, whereas *C. album* had a maximum height of 120 cm. The relation between yield loss and the period between crop and weed emergence for these simulations was presented Fig. 9.2. It appeared that the relationships between yield loss and the period between crop and weed emergence strongly differed between high and low weed densities. However, if yield loss is plotted versus the relative leaf area as determined 30 days after crop emergence for the same simulation experiments a much better relationship was observed for the different species (Fig. 10.1).

Based on these findings, a new, simple descriptive regression model for early prediction of crop losses by weed competition was developed by Kropff and Spitters (1991). This model was derived from the well-tested hyperbolic yield density model and relates yield loss (*YL*) to relative weed leaf area (L_w expressed as weed leaf area / (crop + weed leaf area)) shortly after crop emergence, using the 'relative damage coefficient' q. This approach was described in Chapter 2 (Eqn 2.18). This descriptive model fitted the data accurately and the estimated value for q based on simulated data was close to the observed value for *C. album* (Table 10.1). These results demonstrate the strong impact of maximum height on the damage coefficient q, which characterizes the competitive ability of a species relative to the crop (Fig. 10.1; Table 10.1).

Table 10.1. Estimated parameter values (q) using the relative leaf area model (Eqn 2.18) to fit observed and simulated data sets on competition between *C. album* (simulated with three different maximum heights (*H* is 120, 60 and 30 cm, respectively) and sugar beet (crop height was 60 cm). Source: Kropff and Spitters, 1991.

	q
Sugar beet - *C. album* (observed)	12.10 ± 1.94
Sugar beet - *C. album* (simulated)	
H120	9.62 ± 0.72
H60	1.22 ± 0.03
H30	0.06 ± 0.003

As a second example of how the detailed model may help in *ex-ante* analysis of the potential of a simple predictive approach, the model INTERCOM was used to predict yield loss in a number of competition situations in order to test the ability of the relative leaf area model to describe the effect of mixed weed populations on crop yield. The problem with the relative leaf area model for multiple species (Eqn 2.19) is its inability to account for interspecific competition between the weed species, a process that will affect their competitive ability versus the crop. The model INTERCOM, however, is mechanistically based and takes interspecific competition between weeds in a multi-species situation into account.

The parameters for the relative leaf area model (Eqn 2.18) were derived for single weed species in sugar beet using the results of simulation runs for three different weed species at a wide range of densities and periods between crop and weed emergence fitted with Eqn 2.18 (Table 10.1). The first weed species was *C. album*, the second species had all characteristics of *C. album*, but its maximum height was reduced to 60 cm (the maximum height of the crop) and the third weed species had a maximum height of 30 cm. Multi-weed competition effects were simulated by the model INTERCOM and predicted by the multi-species version of the relative leaf area model (Eqn 2.19). The results were analysed by plotting the simulated yield loss, in which interspecific competition between the weeds is accounted for, versus the predicted yield loss, based on the simple (additive) model and parameters for single species competition (Fig. 10.2). The results show that there are no strong deviations from the 1:1 line, indicating that the error made in the additive leaf area model by neglecting competition between the weed species is relatively small. A deviation was only observed at high yield loss situations, if the weeds strongly differed in maximum height.

Based on the *ex-ante* evaluations of the relative leaf area model and the confirmation with actual data (Chapter 2), it can be concluded that the relative leaf area - yield loss model accounts for the effect of weed densities, the period between crop and weed emergence and different flushes of the weeds, within a limited range of time after crop emergence.

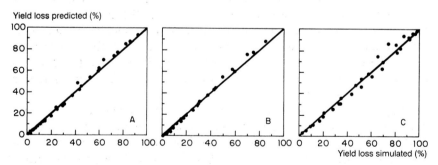

Fig. 10.2. Yield loss caused by a two weed species community simulated with an eco-physiological model composed to predicted yield loss using the regression model (Eqn 2.19) with parameters determined for single species based on simulated data. Weeds have the characteristics of *C. album*, but differ only in their maximum heights. (A) Heights of the two weed species 60 cm; (B) heights of the weeds 30 and 60 cm; (C) heights of the weeds 30 and 120 cm.

To implement the approach in practical decision-making, a methodology has to be developed that enables simple determination of the relative leaf area in the field, e.g. by estimating relative leaf cover with infra-red reflection techniques (Lotz *et al.*, 1993) or the cross-wired sighting device used by Ghersa and Martinez-Ghersa (1991). The potential use of the relative leaf cover model in practical weed management will be increased when interspecific variation in weed life history, morphology and development can be accounted for by lumping weeds into groups of problem weeds. Then, the model could be parameterized per species group (Graf *et al.*, 1990b). Applicability will also be improved by identifying the time window for decision-making with respect to specific problem weeds (Lotz *et al.*, 1992).

Impact of (bio) herbicides on damage relationships

Because of the need to reduce herbicide use, research has started to focus on determining the lowest effective herbicide application. Kudsk (1988, 1989) introduced the concept of 'factor adjusted dose' which accounts for the effect of growth stage of the weeds, weather conditions, etc. In other systems like sugar beet, where herbicide cost forced farmers to find alternatives, experiments with replicated herbicide treatments at a low dosage have been successfully conducted (Wevers, 1991). However, if herbicide dosage is minimized, more weeds will survive the applications. Weeds which escape or survive herbicide applications will have a lower competitive ability compared to untreated weeds, like those used in competition studies. That implies that damage relationships derived from untreated weeds will overestimate competition effects. Weaver (1991) reported a strong reduction in competitive ability of 'escapes' from herbicide treatment (metribuzin). She observed a reduced leaf area for the weeds that were treated. These effects have to be taken into account if damage relationships are used for practical use in combination with low dosages of herbicides.

Figure 10.3 shows the impact of different dosages of a single metamitron treatment on early leaf area development in *C. album* (Lotz and Kropff, unpublished data). The reduction of the relative growth rate of the leaf area is clearly a function of the dosage. Other physiological processes may be affected as well. As shown in Chapter 9, the relative growth rate of the leaf area is the plant trait with most impact on yield loss; i.e. small changes in the relative growth rate of the weeds have a large impact on yield loss. The model INTERCOM can serve as a tool to quantify the impact of these relatively easy to measure physiological/morphological effects on competitive ability of the weeds.

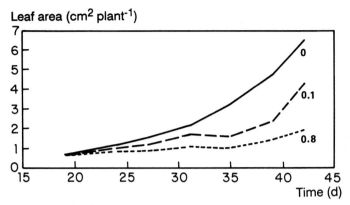

Fig. 10.3. The development of leaf area of *C. album* seedlings with different herbicide treatments: 0, 0.1 and 0.8 kg ha^{-1} (single dosage metamitron, Lotz and Kropff, unpublished results).

Like herbicides, biological control agents are generally also screened for effectiveness in terms of % weed kill. However, most bioherbicides only reduce the growth and development of a weed species, thereby reducing the competitive ability of the weeds (e.g. Wymore and Watson, 1989). If such effects are quantified in simple experiments, the model INTERCOM can be used to predict the impact on yield loss and thus the efficacy of the agents, making laborious competition experiments unnecessary in early stages of agent development (screening).

Spatial weed distribution

One of the first issues raised in discussions about the validity of competition models is the problem related to the spatial heterogeneity of weed distributions in fields. Thornton *et al.* (1990) showed that the spatial distribution of weeds in the field has a substantial effect on the calculation of the economic threshold for weed control. Yield loss - weed density functions should be applied to the different patches separately if weeds are not homogeneously distributed. This is an obvious result of the non-linear yield loss - weed density relationship. An analysis by the model INTERCOM showed that in the sugar beet - *C. album* situation, yield loss per weed m^{-2} was 9.4% at a density of 1 weed plant m^{-2}, but yield loss per weed plant m^{-2} was only 6.5% at 5 weed plants m^{-2}. So, in a patch where the weed density is high, the yield loss caused per weed plant is smaller than when all weeds were distributed homogeneously. Fig. 10.4 shows the impact of clustering of weeds on yield loss. Rice yield loss was simulated at a range of densities and yield loss at average weed density was plotted for a situation without weed-free areas (homogeneous weed distribution), and with a weed-free area of 30 or 60% (weeds clustered on 70 or 40% of the total area). Thornton *et al.* (1990) also demonstrated that the actual distribution in a patch is not of primary importance. This impact of clustering has major implications for economic thresholds for weed control. So, the value of any predictive model for yield loss will be increased if it is applied for the patches that differ in weed infestation in the field separately.

Predicting yield loss in future crops

Weed management systems that are based on the threshold concept could be strongly improved if effects of weeds that are left uncontrolled or are controlled partially on future crops could be predicted. This involves the extension of the competition models to enable simulation of the population dynamics of the weeds. Competition plays an important role in four of the stages of the life cycle of weeds (Fig. 10.5). However, major problems exist with respect to quantification of below ground processes. Field-scale determination of content and dynamics of the seed bank is very labour intensive (Dessaint *et al.*, 1990; Dessaint, 1990) and, therefore, probably not a useful tool in

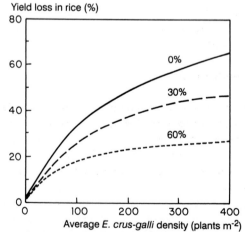

Fig. 10.4. Percentage yield loss as a function of average weed density for a homogeneous weed distribution and a patchy distribution at two levels indicated by a weed-free area of 30% and 60%, as simulated by the model INTERCOM for *E. crus-galli* in direct-seeded rice.

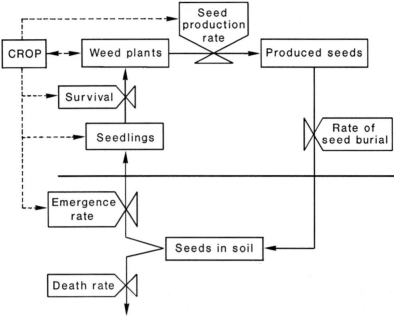

Fig. 10.5. Schematic representation of a population dynamics model for weeds. Broken arrows indicate processes where crop and weeds interact. Source: Kropff and Lotz, 1992.

predicting effects of weed reproduction on yield reductions in subsequent crops. Because of such difficulties, there is a lack of complete data sets to develop and evaluate models for the population dynamics of weeds (Spitters, 1989).

The most pragmatic approach will be the quantification of seed production (or other propagules) per weed plant using competition models (Kropff and Lotz, 1992). Weed management should be directed to the minimization (within economically determined margins) of both yield loss and weed reproductive output. In this respect, it is essential to avoid detailed studies on seed production of each weed species at all relevant environmental conditions. For annual weeds, we can ask whether for these species simple relationships between total biomass of vegetative parts and reproduction do exist. Samson and Werk (1986) developed such a simple model based on a linear relationship between absolute reproductive biomass and vegetative biomass per plant. Thompson *et al.* (1991) demonstrated that in five species of agricultural weeds this linear relationship could be used. Given the validity of this simple relationship, the aforementioned mechanistic simulation models for crop-weed competition offer a powerful tool to predict weed reproduction over a variety of environments, especially when these environments differ in level of competition. Field studies should be initiated to determine the practical applicability of this implementation of weed reproduction in models for crop-weed interactions.

The critical period for weed control

The concept of a critical period for weed control was introduced by Nieto *et al.* (1968). This critical period represents the time interval between two separately determined components: (*i*) the maximum weed-infested period, or the length of time that weeds which emerge with the crop can remain uncontrolled before they begin to compete with the crop and cause yield loss, and (*ii*) the minimum weed-free period, or the length of time that the crop must be free of weeds after sowing, in order to prevent yield losses. These two components are determined in experiments where crop yield loss is measured as a function of successive times of weed removal or emergence, respectively. The weeds in the first component are weeds that emerge more or less simultaneously with the crop, whereas the weeds in the second component are weeds that emerge later than the crop. So, basically the results of two completely different competition situations are combined. It may well be that early weeds can only be tolerated for a very short period because they remove soil-stored resources like water or nutrients at a relatively high rate. In such situations it would not be appropriate to say that the plants actively compete during the critical period. Therefore, it may be better to use the term critical period for weed control instead of weed competition. The results of numerous studies on the critical period in a wide range of crops have been reviewed by Zimdahl (1980, 1988), Radosevich and Holt (1984), and van Heemst (1985).

The use of 'period thresholds' in integrated weed management systems to predict when, rather than if, weeds must be controlled to prevent yield losses was proposed by Dawson (1986). Economic period thresholds could also be calculated, indicating the length of time that a crop could tolerate weed competition before yield loss exceeded the cost of control. This would result in early-season thresholds which would denote the beginning of the critical period, and late-season thresholds the end. Van Heemst (1985) demonstrated that the end of the critical period is closely related to the competitive ability of the crop. Thus, a crop with a high competitive ability has a critical period that ends early.

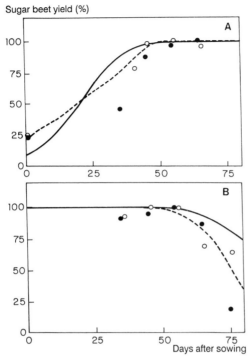

Fig. 10.6. Simulated and observed effect of different weeding treatments on sugar beet yield for 1982 (solid line, o) and 1983 (broken line, •) with (A) different weed-free or (B) weed-infested periods. Yields are presented as percentages of the weed-free controls. Source: Weaver *et al.*, 1992.

The critical period is generally determined by using multiple comparison tests. Cousens (1988, 1991) has suggested using fitted response curves to determine these thresholds. That allows a more accurate analysis. The parameters of the response curves will vary with crop and weed species, weed density, and environmental conditions. Eco-physiological simulation models like INTERCOM may help to analyse how such factors affect the length of the critical period (Weaver *et al.*, 1992).

The model INTERCOM was used to analyse the results of critical period studies for sugar beet and tomatoes. Data on the critical period of *C. album* competition in sugar beet and a complete account of experimental methods were originally published by de Groot and Groeneveld (1986).

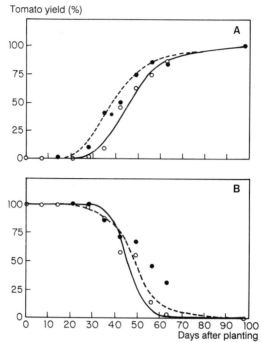

Fig 10.7. Simulated and observed yields of seeded tomatoes for 1981 (solid line, o) and 1982 (broken line, •) with (A) different weed-free or (B) weed-infested periods. Yields are presented as percentages of the weed-free controls. Source: Weaver *et al.*, 1992.

The results of simulation runs for various durations of *C. album* competition and observed data for 1982 and 1983 are shown in Fig. 10.6. There was generally good agreement between simulated and observed yields (linear regression of observed versus simulated yield loss: intercept = −0.2, slope = 0.92; $P < 0.001$).

A similar analysis was conducted for seeded and transplanted tomatoes and a naturally emerging weed population, dominated by *C. album* (Fig. 10.7; Weaver *et al.*, 1992). The impact of weeds in the weed infested series was slightly underestimated for sugar beets and transplanted tomatoes (Weaver *et al.*, 1992). It was hypothesized that the weeds removed nutrients and that the crop needed to express full yield potential in that environment, an effect which is not accounted for in the model. This aspect is not evident in situations where the weeds remain in the crop, as the crop's demand for nutrients remains lower when the weeds are not removed.

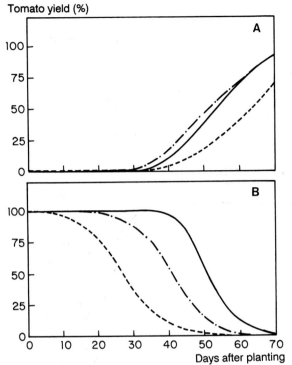

Fig. 10.8. Simulated yields of seeded tomatoes with weather data of 1981 and initial soil moisture of 100 mm as input (solid line), weather data of 1982 and initial soil moisture content of 100 mm as input (dashed line), and weather data of 1982 and initial soil moisture content of 200 mm as input (broken/dotted line) for (A) different weed-free or (B) weed-infested periods. Weed densities were 100 plants m^{-2} for all runs. Source: Weaver *et al.*, 1992.

The model was used to analyse the impact of water limitation and weed density on the weed-free and weed-infested curves.

Water stress strongly affected the weed-infested curves, whereas the weed density (and height, data not shown) had a strong impact on the weed-free curves (Figs 10. 8 and 10.9). These results suggest that in this situation, the period when the crop can tolerate weeds is determined by water competition, whereas the period when the crop has to remain weed free is related more to light competition. These results demonstrate the strength of these eco-physiological models in analysing experimental data and generating information that helps understanding of the system.

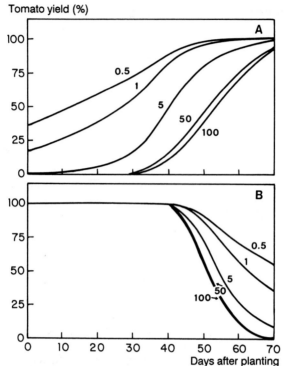

Fig. 10.9. Simulated yields of seeded tomatoes in 1981, with weed densities of 0.5, 1, 5, 50, and 100 plants m⁻² for (A) different weed-free or (B) weed-infested periods. Source: Weaver *et al.*, 1992.

Perspectives for breeding of competitive cultivars

The variation in competitive ability of different crops is large. Van Heemst (1985) compiled data on yield loss in weed-infested crops and ranked the different crop species. Yield loss ranged from 25% in wheat to 49% in transplanted rice to 77% in sugar beet and 100% in onions. Competitive differences between cultivars of a species have been reported as well. However, competitive ability is often negatively correlated to yield potential (e.g. for rice: Jennings and Aquino, 1967; Moody and De Datta, 1982). Wall (1983) reviewed the role of plant breeding in weed management for several crops like rice, wheat and small grain cereals. In general, a high competitive ability was associated with tall plants that rapidly establish complete ground cover. This is in agreement with the results of the analysis of the importance of plant traits for their competitive ability using the

INTERCOM model (Fig. 9.4). Experimentally it is very labour intensive to analyse the competitive ability of a wide range of cultivars, because they have to be tested in a competitive situation (Wall, 1983). By using an eco-physiological model, however, the impact of relatively easy to measure plant traits can be integrated and genetic variability in these traits can be translated into variability in competitive ability for a relevant range of conditions. Lotz *et al.* (1991) applied the model to select sugar beet cultivars with a relative high competitiveness due to early ground cover. They experimentally confirmed strong differences in survival of late emerging weeds between varieties.

The most important trait determining competitive ability was the relative growth rate of the leaf area early in the season (Fig. 9.4), which has to be measured on isolated plants. Spitters and Kramer (1986) measured the relative growth rate of different wheat cultivars (on a dry matter basis) and found a very small coefficient of variation (5%; values ranges from 0.178 to 2.040). They estimated from literature data that the coefficient of variation would range from 3 to 10%. Such a variation would lead to considerable differences in yield loss (Fig. 9.4).

The effect of a 5% increase of the relative growth rate of the leaf area (R_1), and the height growth parameters s and H_{max} of the crop was analysed by the model INTERCOM for rice - *E. crus-galli*. Yield loss at 700 rice plants m^{-2} and 100 *E. crus-galli* plants m^{-2} was 30%. A 5% higher R_1, s or H_{max} in rice reduced the yield loss 2.8, 1.4 and 2.7%, respectively. The combination of improvements led to a 6.4% lower yield loss. Because the genetic variation in rice for these traits is large (Sarkarung, IRRI, personal communication), there may be potential for breeding for higher competitive ability in rice.

In conclusion, it can be stated based on the sensitivity analysis presented in Chapter 9 that morphological characteristics that lead to early ground cover and height development are the most important traits for competitive ability.

A completely different type of application of the model INTERCOM for breeding purposes was described by Kropff *et al.* (1993a). Based on the observation that the position of panicles in rice canopies strongly varies among cultivars, the effect of the position of the panicle through reduction of light absorption by the leaves was studied by the model. The Panicle Area Index can easily reach values of 0.6 - 0.9 m^2 panicle m^{-2} ground. It was shown by the model that panicles in the top 10 cm of the canopy reduce canopy photosynthesis $(LAI = 4)$ by 25%, whereas panicles that are positioned at 20 cm below the top of the leaves reduce canopy photosynthesis by only 10%. Since large genetic variability in panicle height in the canopy exists, this may be a promising characteristic for breeders to increase yield potential as a result of increased light use efficiency, because the panicle surface has a much lower efficiency of using high levels of radiation in photosynthesis than leaves.

Improving weed management

In most agricultural systems, effective weed control has been one of the major problems. The introduction of herbicides was one of the main factors enabling intensification of agriculture in developed countries in past decades. Recently the availability of herbicides has been coupled to intensification of agriculture in developing countries as well. Another important effect of herbicide introduction is the recent area expansion of direct-seeded rice in Asian countries; a technology not widely practised before the late seventies largely because of weed control problems (De Datta and Flinn, 1986).

However, increasing herbicide resistance in weeds, the necessity to reduce cost of inputs (important for some crops like sugar beet) and widespread concern about environmental side effects of herbicides have resulted in greater pressure on farmers to reduce the use of herbicides. This has resulted in the development of strategies for integrated weed management, based on the use of alternative methods for weed control and rationalization of herbicide use, i.e. rather than trying to eradicate weeds from a field, emphasis is on the management of weed populations (Aarts and de Visser, 1985; Baandrup and Ballegaard, 1990; Gerowitt, 1992). The development of such weed management systems requires thorough quantitative insight into the behaviour of weeds in agro-ecosystems and their effects. This involves insight into crop-weed interactions within the growing season as well as insight in the dynamics of weed populations across growing seasons.

The behaviour of weeds in agroecosystems and their effects on crop yield need to be quantified to develop integrated weed management systems. The question to be addressed here is: Can the quantitative tools that have been developed be used to improve weed management systems?

Besides the improvement of weed control technology, two other strategies can be followed to improve weed management systems: (*i*) reduction of weed effects through adapted general crop management practices, and (*ii*) improvement of decision-making with respect to post-emergence weed control measures by predicting yield loss due to weeds at an early stage in the growing season.

With respect to the first strategy, it is important to know what management practices could be followed to favour crops in competition with weeds and to know which plant characteristics, leading to competitive success, might serve as objectives in plant breeding and crop and cultivar selection for different cropping systems. For breeders, however, it will be important to quantify the trade-off between competitive ability and yield potential of the cultivar. Issues like these require thorough quantitative insight into the competition process, and how physiological and morphological growth characteristics of weeds and the crop are related to competitive relationships between crop and weed (Berkowitz, 1988). Systems analysis and simulation may help to bridge the gap between knowledge at the process level and management at the field level. As discussed earlier in this chapter, the simulation model described in this book can be used to identify traits that determine the competitiveness of the crop. The INTERCOM model could also be a useful tool to analyse yield trials of breeders, to determine the trade-off between such traits and yield potential.

Since weed problems obviously cannot be solved by adaptation of general crop management practices alone, insight into the decision-making process of farmers is needed to determine what knowledge should be available and in what form. The decision-making process in weed management based on post-emergence observations is illustrated in Fig. 10.10.

To allow rational decision-making, the severity of weed infestation shortly after crop emergence should be quantified using a simple practical method. Weed effects on crop yield have to be predicted accurately on the basis of these observations. These observations should be repeated until newly emerging weeds no longer affect crop yield. The efficacy and cost of different possible weed control measures (mechanical, chemical or biological) also have to be quantified on the basis of observations of the weed infestation. Criteria must be defined (i.e., the cost effectiveness of weed control) to enable economic decision-making.

Thus, a simple measure for estimating the severity of weed infestation has to be defined that allows accurate prediction of weed effects at the end of the growing season. Besides that, cost, efficacy and side effects of possible control measures have to be quantified. The work of Auld *et al.* (1987) on economic aspects could be used for that purpose. Such information can be used to decide whether and how the weeds should be controlled using well-defined criteria such as maximization of profits and minimization of environmental effects. Until now, such approaches to weed management have scarcely been tested (Cousens, 1987).

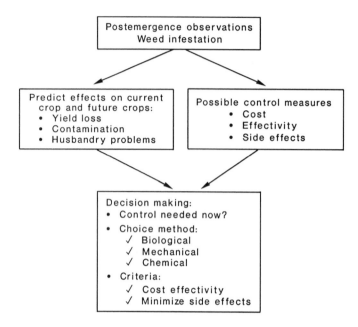

Fig. 10.10. A scheme for decision-making in integrated weed management systems. Source: Kropff and Lotz, 1992.

Discussion and conclusions

Eco-physiological simulation models for crop-weed interactions can well be used to link field-level observations of weed effects to underlying physiological and morphological processes, and thus provide understanding. They can also be helpful in the design of effective simple predictive regression models which can be used at the farm level. An important aspect is that the same regression model that is used in weed management systems is also used by the scientist who observes yield loss caused by the weeds. Therefore, it is essential to develop a sound approach that has the potential for practical application, before starting laborious experiments. This way of linking research at different levels and practical application is illustrated in Fig. 10.11.

The use of systems approaches can encourage weed ecologists to produce challenging questions for weed technologists. An example is the separation of the effects of weeds in current and future crops. Concepts that have been developed for pest management in relationship to tactical (within season) and strategic (longer time frame) pest management (Rabbinge and Rossing, 1987) could be very useful in

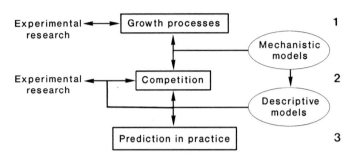

Fig. 10.11. Schematic indication of the potentials to use descriptive (regression) models and mechanistic (eco-physiological) models in research and practice. Source: Kropff and Lotz, 1992.

developing weed management systems. Often weeds do not cause yield loss in a current crop, for example late-emerging weeds, weeds in winter wheat growing under favourable conditions (Lotz *et al.*, 1990). In such situations, we need new technology to avoid or reduce weed seed production (cf. Medd and Ridings, 1989) and analyses of the economical aspects (Pandey and Medd, 1990). Besides new chemical technologies, biological knowledge and insights could be used to develop technologies that allow interference with the development of plants. An idea could be to prevent flowering in short-day plants (weeds) when days become shorter by interrupting the night period using light flashes.

When research is based on sound insight into the system, it should be possible to develop and optimize weed management systems that are as effective as current weed management procedures, while meeting other criteria such as cost effectiveness and minimization of environmental effects. Such effective weed management systems should combine preventive practices, crop management practices, cultural practices and the range of weed control technologies (Moody, 1991).

Until now modelling efforts in weed science have concentrated on crop-weed competition and population dynamics excluding other critical areas like invasion, rate of spread, effectiveness and economics of weed control (Doyle, 1991). The future challenge for modelling in weed science will be the development and integration of the different components.

References

Aarts, H.F.M. and de Visser, C.L.M. (1985) A management information system for weed control in winter wheat. In: *Proceedings 1985 British Crop Protection Conference* (BCPC) - Weeds, 7A-2, pp. 679-86.

Acock, B., Charles-Edwards, D.A., Fitter, D.J., Hand, D.W., Ludwig, L.J., Warren Wilson, J. and Withers, A.C. (1978) The contribution of leaves from different levels within a tomato crop to canopy net photosynthesis: an experimental examination of two canopy models. *Journal of Experimental Botany* 29, 815-27.

Altieri, M.A. and Liebman, M. (eds) (1988) *Weed Management in Agroecosystems: Ecological Approaches*. CRC Press Inc., Boca Raton, Florida, 354 pp.

Ampong-Nyarko, K. and De Datta, S.K. (1989) Ecophysiological studies in relation to weed management strategies in rice. In: *Proceedings of the Asian-Pacific Weed Science Society*, Twelfth Conference, Seoul, Korea, pp. 31-45.

Amthor, J.S. (1984) The role of maintenance respiration in plant growth. *Plant, Cell and Environment* 7, 561-9.

Angus, J.F. (1991) The evolution of methods for quantifying risk in water limited environments. In: Muchow, R.C. and Bellamy, J.A. (eds) *Climatic Risk in Crop Production: Models and Management for the Semiarid Tropics and Subtropics*. CAB International, Wallingford, UK, pp. 39-53.

Angus, J.F., Cunningham, R.B., Moncur, M.W. and Mackenzie, D.H. (1981) Phasic development in field crops. I. Thermal response in the seedling phase. *Field Crops Research* 3, 365-78.

Auld, B.A., Menz, K.M. and Tisdell, C.A. (1987) Weed Control Economics. Academic Press, Inc., London, 177 pp.

Baandrup, M. and Ballegaard, M.T. (1990) Advisory computer system for weed control. In: *Proceedings EWRS Symposium Integrated Weed Management in Cereals*, Helsinki, pp. 443-51.

Barnes, P.W., Beyschlag, W., Ryel, R., Flint, S.D. and Caldwell, M.M. (1990) Plant competition for light analyzed with a multispecies canopy model. III. Influence of canopy structure in mixtures and monocultures of wheat and wild oat. *Oecologia* 82, 560-6.

Barnett, K.H. and Pearce, R.B. (1983) Source-sink ratio alteration and its effect on physiological parameters in maize. *Crop Science* 23, 294-9.

Bazzaz, F.A., Garbutt, K., Reekie, E.G. and Williams, W.E. (1989) Using growth analysis to interpret competition between C_3 and C_4 annuals under ambient and elevated CO_2. *Oecologia* 79, 223-35.

Berendse, F., Oudhof, H. and Bol, J. (1987) A comparative study on nutrient cycling in wet heathlands ecosystems. I. Litter production and nutrient losses from the plant. *Oecologia* 74, 174-84.

Berendse, F., Bobbink, R. and Rouwenhorst, G. (1989) A comparative study on nutrient cycling in wet heathlands ecosystems. II. Litter decomposition and nutrient mineralization. *Oecologia* 78, 338-48.

Berkowitz, A.R. (1988) Competition for resources in weed-crop mixtures. In: Altieri, M.A. and Liebman, M. (eds) *Weed Management in Agroecosystems: Ecological Approaches.* CRC Press Inc., Boca Raton, Florida, pp. 89-119.

Beyschlag, W., Barnes, P.W., Ryel, R., Caldwell, M.M. and Flint, S.D. (1990) Plant competition for light analyzed with a multispecies canopy model. II. Influence of photosynthetic characteristics on mixtures of wheat and wild oat. *Oecologia* 82, 374-80.

Black, C.C., Chen, T.M. and Brown, R.H. (1969) Biochemical basis for plant competition. *Weed Science* 17, 338-44.

Brown, R.H. (1985) Growth of C_3 and C_4 grasses under low N levels. *Crop Science* 25, 954-7.

Brunt, D. (1932) Notes on radiation in the atmosphere. I. *Quarterly Journal of the Royal Meteorological Society* 58, 389-420.

Caldwell, R. and Hansen, J. (1990) Simulation of intercropping with IBSNAT models. *Agrotechnology Transfer* 12, 1-4.

Carberry, P.S. (1991) Test of leaf-area development in CERES-Maize: a correction. *Field Crops Research* 27, 159-67.

Chu, Chang Chi, Ludford, P.M., Ozbun, J.L. and Sweet, R.D. (1978) Effects of temperature and competition on the establishment of redwood pigweed and common *Chenopodium album* (L.). *Crop Science* 18, 308-11.

Clark, E.A. and Loomis, R.S. (1978) Dynamics of leaf growth and development in sugar beets. *Journal of the American Society of Sugar Beet Technologists* 20, 97-113.

Connolly, J. (1986) On difficulties with replacement-series methodology in mixture experiments. *Journal of Applied Ecology* 23, 125-37.

Cousens, R. (1985) An empirical model relating crop yield to weed and crop density and a statistical comparison with other models. *Journal of Agricultural Science* 105, 513-21.

Cousens, R. (1987) Theory and reality of weed control thresholds. *Plant Protection Quarterly* 2(1), 13-20.

Cousens, R. (1988) Misinterpretations of results in weed research through inappropriate use of statistics. *Weed Research* 28, 281-9.

Cousens, R. (1991) Aspects of the design and interpretation of competition (interference) experiments. *Weed Technology* 5(3), 664-73.

Cousens, R., Brain, P., O'Donovan, J.T. and O'Sullivan, A. (1987) The use of biologically realistic equations to describe the effects of weed density and relative time of emergence on crop yield. *Weed Science* 35, 720-5.

Darwin, C. (1859) *The Origin of Species by Means of Natural Selection or the Preservation of Favoured Races in the Struggle for Life.* Murray, London.

Dawson, J.H. (1986) The concept of period thresholds. In: *Proceedings of the European Weed Research Society.* Symposium on Economic Weed Control, Stuttgart-Hohenheim, Germany, pp. 327-31.

De Datta, S.K. and Flinn, J.C. (1986) Technology and economics of weed control in broadcast-seeded flooded tropical rice. In: Noda, K. and Mercado, B.L. (eds) *Weeds and the Environment in the Tropics.* Proceedings of the 10th Asian-Pacific Weed Science Society Conference, Chiangmai, Thailand, pp. 51-74.

Dessaint, F. (1990) La répartition spatiale du stock semencier: comparaison de techniques statistiques. *Weed Research* 31, 41-8.

Dessaint, F., Barralis, G., Beuret, E., Caixinhas, M.L., Post, B.J. and Zanin, G. (1990) Etude coopérative EWRS: la détermination du potentiel semencier: I. Recherche d'une relation entre la moyenne et la variance déchantillonnage. *Weed Research* 30, 421-9.

Dobben, W.H. van (1962) Influence of temperature and light conditions on dry matter distribution, development rate and yield in arable crops. *Netherlands Journal of Agricultural Science* 10, 377-89.

Doorenbos, J. and Kassam, A.H. (1979) *Yield Response to Water.* FAO Irrigation and Drainage Paper 33, FAO, Rome, 193 pp.

Doyle, C.J. (1991) Mathematical models in weed management. *Crop Protection* 10(6), 432-44.

Driessen, P.M. (1982) Nutrient demand and fertilizer requirements. In: van Keulen, H. and Wolf, J. (eds) *Modelling of Agricultural Production: Weather, Soils and Crops.* Simulation Monographs, Pudoc, Wageningen, pp. 182-200.

Ehleringer, J. and Pearcy, R.W. (1983) Variation in quantum yield for CO_2 uptake among C_3 and C_4 plants. *Plant Physiology* 73, 555-9.

Feddes, R.A. (1987) Crop factors in relation to Makkink reference-crop evapotranspiration. TNO Committee on Hydrological Research, The Hague, The Netherlands. *Proceedings and Information* 39, 33-46.

Firbank, L.G. and Watkinson, A.R. (1985) On the analysis of competition within two-species mixtures of plants. *Journal of Applied Ecology* 22, 503-17.

Firbank, L.G. and Watkinson, A.R. (1987) On the analysis of competition at the level of the individual plant. *Oecologia* 71, 308-17.

Firbank, L.G. and Watkinson, A.R. (1990) On the effects of competition: from monocultures to mixtures. In: Grace, J.B. and Tilman, D. (eds) *Perspectives on Plant Competition*. Academic Press, Inc., San Diego, pp. 165-92.

Frère, M. and Popov, G.F. (1979) *Agrometeorological Crop Monitoring and Forecasting*. Plant Production and Protection Paper 17, FAO, Rome, 64 pp.

Frissel, M.J. and van Veen, J.A. (eds) (1981) *Simulation of Nitrogen Behaviour of Soil-Plant Systems*. Pudoc, Wageningen, 277 pp.

Fukuda, H. and Hayashi, I. (1982) Ecology of dominant plant species of early stages in secondary succession, on *Chenopodium album* (L.). *Japanese Journal of Ecology* 32, 517-26.

Gerowitt, B. (1992) Dreijährige Versuche zur Anwendung eines computergestützten Entscheidungsmodel zur Unkrautbekämpfung nach Schadenschwellen in Winterweizen. *Zeitschrift für Pflanzenkrankheiten und Pflanzenschutz* 13, 301-10.

Gerwen, C.P. van, Spitters, C.J.T. and Mohren, G.M.J. (1987) Simulation of competition for light in even-aged stands of Douglas fir. *Forest Ecology and Management* 18, 135-52.

Ghersa, C.M. and Martinez-Ghersa, M.A. (1991) A field method for predicting yield losses in maize caused by Johnsongrass (*Sorghum halepense*). *Weed Technology* 5, 279-85.

Gijzen, H. and Goudriaan, J. (1989) A flexible and explanatory model of light distribution and photosynthesis in row crops. *Agricultural and Forest Meteorology* 48, 1-20.

Gosse, G., Varlet-Grancher, C., Bonhomme, R., Chartier, M., Allirand, J.-M. and Lemaire, G. (1986) Production maximale de matière sèche et rayonnement solaire intercepté par un couvert végétal. *Agronomie* 6, 47-56.

Goudriaan, J. (1977) *Crop Micrometeorology: a Simulation Study*. Simulation Monographs, Pudoc, Wageningen, 257 pp.

Goudriaan, J. (1982) Some techniques in dynamic simulation. In: Penning de Vries, F.W.T. and van Laar, H.H. (eds) *Simulation of Plant Growth and Crop Production*. Simulation Monographs, Pudoc, Wageningen, pp. 66-84.

Goudriaan, J. (1986) A simple and fast numerical method for the computation of daily totals of canopy photosynthesis. *Agricultural and Forest Meteorology* 43, 251-5.

Goudriaan, J. (1988) The bare bones of leaf-angle distribution in radiation models for canopy photosynthesis and energy exchange. *Agricultural and Forest Meteorology* 43, 155-69.

Goudriaan, J. and van Keulen, H. (1979) The direct and indirect effects of nitrogen shortage on photosynthesis and transpiration in maize and sunflower. *Netherlands Journal of Agricultural Science* 27, 227-34.

Grace, J.B. (1990) On the relationship between plant traits and competitive ability. In: Grace, J.B. and Tilman, D. (eds) *Perspectives on Plant Competition*. Academic Press, Inc., San Diego, pp. 51-65.

Grace, J.B. and Tilman, D. (eds) (1990) *Perspectives on Plant Competition*. Academic Press, Inc., San Diego, 484 pp.

Graf, B. and Hill, J.E. (1992) Modelling the competition for light and nitrogen between rice and *Echinochloa crus-galli*. *Agricultural Systems* 40, 345-59.

Graf, B., Rakotobe, O., Zahner, P., Delucchi, V. and Gutierrez, A.P. (1990a) A simulation model for the dynamics of rice growth and development: Part I. The carbon balance. *Agricultural Systems* 32, 341-65.

Graf, B., Gutierrez, A.P., Rakotobe, O., Zahner, P. and Delucchi, V. (1990b) A simulation model for the dynamics of rice growth and development. Part II. The competition with weeds for nitrogen and light. *Agricultural Systems* 32, 367-92.

Greenwood, D.J., Draycott, A., Last, P.J. and Draycott, A.P. (1984) A concise simulation model for interpreting N-fertilizer trials. *Fertilizer Research* 5, 355-69.

Grime, J.P. (1979) *Plant Strategies and Vegetation Processes*. Wiley, Chichester, 222 pp.

Groot, J.J.R., de Willigen, P. and Verberne, E.L.J. (eds) (1991) *Nitrogen Turnover in the Soil-Crop System*. Special Issue of *Fertilizer Research* 27(2-3), 141-383.

Groot, W. de and Groeneveld, R.M.W. (1986) Weed control in sugar beet in relation to the onset of the critical period of weed competition. *Gewasbescherming* 17, 171-8. (in Dutch)

Håkansson, S. (1983) *Competition and Production in Short-Lived Crop-Weed Stands*. Sveriges Landbruks University, Uppsala, Sweden. Report 127, 85 pp.

Hall, A.E. and Loomis, R.S. (1972) Photosynthesis and respiration by healthy and beet yellows virus infected sugar beets (*Beta vulgaris* L.). *Crop Science* 12, 566-71.

Harper, J.L. (1977) *The Population Biology of Plants*. Academic Press,

London, 892 pp.

Heemst, H.D.J. van (1985) The influence of weed competition on crop yield. *Agricultural Systems* 18, 81-93.

Hill, J.E., De Datta, S.K. and Real, J.G. (1990) *Echinochloa* competition in rice: a comparison of studies from direct-seeded and transplanted flooded rice. In: Auld, B.A., Umaly, R.C. and Tjitrosomo, S.S. (eds) *Proceedings of the Symposium on Weed Management*. Biotrop Special Publication No. 38, Bogor, Indonesia, pp. 115-29.

Hodanova, D. (1981) Photosynthetic capacity, irradiance and sequential senescence of sugar beet leaves. *Biologia Plantarum* 23, 58-67.

Horie, T., de Wit, C.T., Goudriaan, J. and Bensink, J. (1979) A formal template for the development of cucumber in its vegetative stage. *Proceedings of the Koninklijke Nederlandse Akademie van Wetenschappen* Series C, Volume 82 (4), 433-79.

IRRI (1989) *IRRI Toward 2000 and Beyond*. International Rice Research Institute, P.O. Box 933, 1099 Manila, Philippines, 66 pp.

Jansen, D.M. (1990) Potential rice yields in future weather conditions in different parts of Asia. *Netherlands Journal of Agricultural Science* 38, 661-80.

Janssen, B.H., Lathwell, D.J. and de Wit, C.T. (1987) Modeling long-term crop response to fertilizer phosphorus. II. Comparison with field results. *Agronomy Journal* 79, 452-8.

Janssen, B.H., Guiking, F.C.T., van der Eijk, D., Smaling, E.M.A., Wolf, J. and van Reuler, H. (1990) A system for quantifying evaluation of the fertility of tropical soils (QUEFTS). *Geoderma* 46, 299-318.

Jennings, P.R. and Aquino, R.C. (1967) Studies on competition in rice. III. The mechanism of competition among phenotypes. *Evolution* 22, 529-42.

Jones, C.A. and Kiniry, J.R. (1986) *CERES-Maize: a Simulation Model of Maize Growth and Development*. Texas A&M University Press, College Station, Texas.

Jones, J.W. and Hesketh, J.D. (1980) Predicting leaf expansion. In: Hesketh, J.D. and Jones, J.W. (eds) *Predicting Photosynthesis for Ecosystem Models*. CRC Press Inc., Boca Raton, Florida, Vol. 2, pp. 85-122.

Jones, J.W., Boote, K.J., Hoogenboom, G., Japtap, S.S. and Wilkerson, C.G. (1989) *SOYGRO V5.42, Soybean Crop Growth Simulation Model: User's Guide*. Florida Agricultural Experiment Station Journal No. 8304. Agricultural Engineering and Agronomy Department, University of Florida, Gainsville, Florida.

Keulen, H. van (1975) *Simulation of Water Use and Herbage Growth in Arid Regions*. Simulation Monographs, Pudoc, Wageningen,

176 pp.

Keulen, H. van (1982) Crop production under semi-arid conditions, as determined by nitrogen and moisture availability. In: Penning de Vries, F.W.T. and van Laar, H.H. (eds) *Simulation of Plant Growth and Crop Production*. Simulation Monographs, Pudoc, Wageningen, pp. 234-49.

Keulen, H. van (1986) Crop yield and nutrient requirements. In: van Keulen, H. and Wolf, J. (eds) *Modelling of Agricultural Production: Weather, Soils and Crops*. Simulation Monographs, Pudoc, Wageningen, pp. 153-81.

Keulen, H. van and Louwerse, W. (1975) Simulation models for plant production. In: *Proceedings WMO Symposium 'On Agrometeorology of the Wheat Crop'*. Braunschweig, 1973. Deutscher Wetterdienst, Offenbach am Mainz, pp. 196-209.

Keulen, H. van and Wolf, J. (eds) (1986) *Modelling of Agricultural Production: Weather, Soils and Crops*. Simulation Monographs, Pudoc, Wageningen, 479 pp.

Keulen, H. van and Seligman, N.G. (1987) *Simulation of Water Use, Nitrogen and Growth of a Spring Wheat Crop*. Simulation Monographs, Pudoc, Wageningen, 310 pp.

Keulen, H. van, Penning de Vries, F.W.T. and Drees, E.M. (1982) A summary model for crop growth. In: Penning de Vries, F.W.T. and van Laar, H.H. (eds) *Simulation of Plant Growth and Crop Production*. Simulation Monographs, Pudoc, Wageningen, pp. 87-97.

Kira, T., Ogawa, H. and Sakazaki, N. (1953) Intraspecific competition among higher plants. I. Competition-yield-density interrelationships in regularly dispersed populations. *Journal of Institute for Polytechnics, Osaka City University*, Series D 4, 1-16.

Koster, P.B., Raats, P. and Jorritsma, J. (1980) *The Effect of Some Agronomical Factors on the Respiration Rates of Sugar Beet*. Instituut voor Rationele Suikerproductie, Bergen op Zoom, the Netherlands, Mededeling nr. 6.

Kraalingen, D.W.G. van (1989) *A Three-Dimensional Light Model for Crop Canopies*. Internal Report 17, Department of Theoretical Production Ecology, Wageningen Agricultural University, 35 pp.

Kraalingen, D.W.G. van (1991) *The FSE System for Crop Simulation*. Simulation Report CABO-TT no. 23, Centre for Agrobiological Research, P.O. Box 14, 6700 AA Wageningen, 77 pp.

Kraalingen D.W.G. van, Stol, W., Uithol, P.W.J. and Verbeek, M. (1991) *User Manual of CABO/TPE Weather System*. CABO/TPE Internal communication, Centre for Agrobiological Research, P.O. Box 14, 6700 AA Wageningen, 27 pp.

Kropff, M.J. (1988a) Modelling the effects of weeds on crop production. *Weed Research* 28, 465-71.

Kropff, M.J. (1988b) Simulation of crop-weed competition. In: Miglietta, F. (ed.) *Models in Agriculture and Forest Research*. Proceedings of a workshop held at San Miniato, June 1-3 1987, Italy, pp. 177-86.

Kropff, M.J. (1988c) Simulation of crop-weed competition. In: Cavelloro, R. and El Titi, A. (eds) *Weed Control in Vegetable Production*. Proceedings of a meeting of the EC expert's group. Stuttgart, 1986, pp. 73-83.

Kropff, M.J. (1989) *Quantification of SO₂ Effects on Physiological Processes, Plant Growth and Crop Production*. PhD thesis, Department of Theoretical Production Ecology, Wageningen Agricultural University, 201 pp.

Kropff, M.J. (1990) The effects of long-term open-air fumigation with SO_2 on a field crop of broad bean (*Vicia faba* L.). III. Quantitative analysis of damage components. *New Phytologist* 115, 357-65.

Kropff, M.J. and Goudriaan, J. (1989) Modelling short-term effects of sulphur dioxide. 3. Effects of SO_2 on photosynthesis of leaf canopies. *Netherlands Journal of Plant Pathology* 95, 265-80.

Kropff, M.J. and Spitters, C.J.T. (1991) A simple model of crop loss by weed competition from early observations on relative leaf area of the weeds. *Weed Research* 31, 97-105.

Kropff, M.J. and Spitters, C.J.T. (1992) An eco-physiological model for interspecific competition, applied to the influence of *Chenopodium album* L. on sugar beet. I. Model description and parameterization. *Weed Research* 32, 437-50.

Kropff, M.J. and Lotz, L.A.P. (1992) Systems approaches to quantify crop-weed interactions and their application in weed management. *Agricultural Systems* 40, 265-82.

Kropff, M.J., Vossen, F.J.H., Spitters, C.J.T. and de Groot, W. (1984) Competition between a maize crop and a natural population of *Echinochloa crus-galli* (L.). *Netherlands Journal of Agricultural Science* 32, 324-7.

Kropff, M.J., Joenje, W., Bastiaans, L., Habekotté, B., van Oene, H. and Werner, R. (1987) Competition between a sugar beet crop and populations of *Chenopodium album* (L.) and *Stellaria media* (L.). *Netherlands Journal of Agricultural Science* 35, 525-8.

Kropff, M.J., Weaver, S.E. and Smits, M.A. (1992a) Use of ecophysiological models for crop-weed interference: Relations amongst weed density, relative time of weed emergence, relative leaf area, and yield loss. *Weed Science* 40, 296-301.

Kropff, M.J., Spitters, C.J.T., Schnieders, B.J., Joenje, W. and de Groot, W. (1992b) An eco-physiological model for interspecific competition, applied to the influence of *Chenopodium album* L. on sugar beet. II. Model evaluation. *Weed Research* 32, 451-63.

Kropff, M.J., Cassman, K.G. and van Laar, H.H. (1993a) Quantitative understanding of the irrigated rice ecosystem for achieving high yield potential in (hybrid) rice. In: *Proceedings of the International Rice Research Conference*, 21-25 April, 1992. International Rice Research Institute, Los Banos, Philippines (submitted).

Kropff, M.J., van Laar, H.H. and ten Berge, H.F.M. (eds) (1993b) *ORYZA1: a Basic Model for Irrigated Lowland Rice Production.* International Rice Research Institute, Los Banos, Philippines, 89 pp.

Ku, M.S.B., Schmitt, M.R. and Edwards, G.E. (1979) Quantitative determination of RuBP carboxylase - oxygenase protein in leaves of several C_3 and C_4 plants. *Journal of Experimental Botany* 30, 89-98.

Kudsk, P. (1988) The influence of volume rates on the activity of glycophosate and difenzoquat assessed by a parallel-line assay technique. *Pesticide Science* 24, 21-9.

Kudsk, P. (1989) Experiences with reduced herbicide doses in Denmark and the development of the concept of factor-adjusted doses. In: *Proceedings of the Crop Protection Conference - Weeds*, Brighton, pp. 545-54.

Leffelaar, P.A. and Ferrari, Th.J. (1989) Some elements of dynamic simulation. In: Rabbinge, R., Ward, S.A. and van Laar, H.H. (eds) *Simulation and Systems Management in Crop Protection.* Simulation Monographs, Pudoc, Wageningen, pp. 19-45.

Légère, A. and Schreiber, M.M. (1989) Competition and canopy architecture as affected by soybean (*Glycine max*), row width and density of redroot pigweed (*Amaranthus retroflexus*). *Weed Science* 37, 84-92.

Lotz, L.A.P., Kropff, M.J. and Groeneveld, R.M.W. (1990) Herbicide application in winter wheat. Experimental results on weed competition analyzed by a mechanistic model. *Netherlands Journal of Agricultural Science* 30, 711-8.

Lotz, L.A.P., Groeneveld, R.M.W. and de Groot, N.A.M.A. (1991) Potential for reducing herbicide inputs in sugar beet by selecting early closing cultivars. In: *Proceedings Crop Protection Conference - Weeds*, Brighton, 9A-8, pp. 1241-8.

Lotz, L.A.P., Kropff, M.J., Bos, B. and Wallinga, J. (1992) Prediction of yield loss based on relative leaf cover of weeds. In: Richardson, R.G. (compiler) *Proceedings of the First International Weed Control Congress*, Volume 2, Monash University, Melbourne, Australia, pp. 290-6.

Lotz, L.A.P., Kropff, M.J., Wallinga, J. Bos, H.J., Groeneveld, R.M.W. (1993) Techniques to estimate relative leaf area and cover of weeds in crops for yield loss prediction. *Weed Research* (in press).

Louwerse, W., Sibma, L. and van Kleef, J. (1990) Crop photosynthesis, respiration and dry matter production of maize. *Netherlands Journal of Agricultural Science* 38, 95-108.

Lubigan, R.T. and Vega, M.R. (1971) The effect on yield of competition of rice with *Echinochloa crus-galli* (L.) Beauv. and *Monochoria vaginalis* (Burm. f.) Presl. *Philippine Agriculturist* 55, 210-3.

Medd, R.W. and Ridings, H.I. (1989) Relevance of seed kill for the control of annual grass weeds in crops. In: Delfosse, E.S. (ed.) *Proceedings VII International Symposium on Biological Control of Weeds*, 6-11 March 1988, Rome, pp. 645-50.

Miglietta, F. (1989) Effect of photoperiod and temperature on leaf initiation rates in wheat (*Triticum* spp.). *Field Crops Research* 21, 121-30.

Miglietta, F. (1991a) Simulation of wheat ontogenesis. I. Appearance of main stem leaves in the field. *Climate Research* 1, 145-50.

Miglietta, F. (1991b) Simulation of wheat ontogenesis. II. Predicting dates of ear emergence and main stem final leaf number. *Climate Research* 1, 151-60.

Milford, G.F.J. and Riley, J. (1980) The effects of temperature on leaf growth of sugar beet varieties. *Annals of Applied Biology* 94, 431-43.

Miyasaka, A., Murata, Y. and Iwata, I. (1975) Leaf area development and leaf senescence in relation to climate and other factors. In: Murata, Y. (ed.) *Crop Productivity and Solar Energy Utilization in Various Climates in Japan*. Japanese Committee for the International Biological Program (JIBP), University of Tokyo Press, Volume 11, pp. 72-85.

Monsi, M. and Saeki, T. (1953) Ueber den Lichtfaktor in den Pflanzengesellschaften und sein Bedeutung für die Stoffproduktion. *Japanese Journal of Botany* 14, 1-22.

Monteith, J.L. (1965) Evaporation and environment. *Proceedings Symposium Society of Experimental Biology* 19, 205-34.

Monteith, J.L. (1969) Light interception and radiative exchange in crop stands. In: Eastin, J.D., Haskins, F.A., Sullivan, C.Y. and van Bavel, C.H.M. (eds) *Physiological Aspects of Crop Yield*. American Society of Agronomy, Crop Science Society of America, Madison, Wisconsin, U.S.A., pp. 89-111.

 Moody, K. (1991) Weed management in rice. In: Pimentel, D. (ed.) *Handbook of Pest Management in Agriculture*. 2nd edition, CRC Press Inc., Boca Raton, Florida, pp. 301-28.

Moody, K. and De Datta, S.K. (1982) Integration of weed control practices for rice in tropical Asia. In: Soerjani, M., Barnes, D.E. and Robson, T.O. (eds) *Weed Control in Small Farms*. Asian-Pacific Weed Science Society, Biotrop Special Publication No. 15,

Bogor, pp. 37-47.

Morgan, D.C. and Smith, H. (1981) Non-photosynthetic responses to light quality. In: Lange, D.L., Nobel, P.S., Osmond, C.B. and Ziegler, H. (eds) *Encyclopedia of Plant Physiology* Volume 12A, Physiological Plant Ecology I, Springer Verlag, Berlin, Heidelberg, New York, pp. 109-34.

Muchow, R.C. and Carberry, P.S. (1989) Environmental control of phenology and leaf growth of maize in a semiarid tropical environment. *Field Crops Research* 20, 221-36.

Ng, E. and Loomis, R.S. (1984) *Simulation of Growth and Yield of the Potato Crop*. Simulation Monographs, Pudoc, Wageningen, 147 pp.

Niemann, P. (1986) Mehrjährige Anwendung des Schadensschwellenprinzips bei der Unkrautbekämpfung auf einem landwirtschaftlichen Betrieb. *Proceedings of the European Weed Research Society Symposium 1986*, Economic Weed Control, Stuttgart-Hohenheim, Germany, pp. 385-92.

Nieto, H.J., Brondo, M.A. and Gonzalez, J.T. (1968) Critical periods of the crop growth cycle for competition from weeds. *Pest Articles and News Summaries* (C) 14, 159-66.

Noda, K. (1973) Competition effects of barnyardgrass (*Echinochloa crus-galli*) on rice. In: *Proceedings of the 4th Asian-Pacific Weed Science Society Conference*, Rotorua, New Zealand, pp. 145-50.

Noda, K., Ozawa, K. and Ibaraki, K. (1968) Studies on the damage on rice plants due to weed competition (effect of barnyardgrass competition on growth, yield, and some eco-physiological aspects of rice plants). *Bulletin Kyushu Agricultural Experimental Station* 13(3-4), 345-67.

Noordwijk, M. van (1983) Functional interpretation of root densities in the field for nutrient and water uptake. In: *Root Ecology and its Practical Application*. International Symposium Gumpenstein, 1982, Bundesanstalt Gumpenstein, Irdning, Austria, pp. 207-26.

Norris, R.F. (1992) Have ecological and biological studies improved weed control strategies? In: Combellack, J.H., Levick, K.J., Parsons, J. and Richardson, R.G. (eds) *Proceedings of the First International Weed Control Congress*, Volume 1, Monash University, Melbourne, Australia, pp. 7-33.

Oorschot, J.L.P. van and van Leeuwen, P.H. (1984) Comparison of photosynthetic capacity between intact leaves of triazine resistant and susceptible biotypes of six weed species. *Zeitschrift für Naturforschung* 39, 440-2.

Pacala, S.W. and Silander Jr, J.A. (1987) Neighborhood interference among velvet leaf, *Abutilon theophrasti*, and pigweed, *Amaranthus retroflexus*. *OIKOS* 48, 217-24. (Copenhagen 1987).

Pandey, S. and Medd, R.W. (1990) Integration of seed and plant kill tactics for control of wild oats: an economic evaluation. *Agricultural Systems* 34, 65-76.

Pearcy, R.W., Tumosa, N. and Williams, K. (1981) Relationships between growth, photosynthesis and competitive interactions for a C_3 and C_4 plant. *Oecologia* 48, 371-6.

Penman, H.L. (1948) Natural evaporation from open water, bare soil and grass. *Proceedings of the Royal Society of London* Series A 193, 120-46.

Penman, H.L. (1956) Evaporation: An introductory survey. *Netherlands Journal of Agricultural Science* 4, 9-29.

Penning de Vries, F.W.T. (1975) The cost of maintenance processes in plant cells. *Annals of Botany* 39, 77-92.

Penning de Vries, F.W.T. (1982) Systems analysis and models of crop growth. In: Penning de Vries, F.W.T. and van Laar, H.H. (eds) *Simulation of Plant Growth and Crop Production*. Simulation Monographs, Pudoc, Wageningen, pp. 9-19.

Penning de Vries, F.W.T and van Laar, H.H. (1982a) Simulation of growth processes and the model BACROS. In: Penning de Vries, F.W.T. and van Laar, H.H. (eds) *Simulation of Plant Growth and Crop Production*. Simulation Monographs, Pudoc, Wageningen, pp. 114-35.

Penning de Vries, F.W.T. and van Laar, H.H. (eds) (1982b) *Simulation of Plant Growth and Crop Production*. Simulation Monographs, Pudoc, Wageningen, 308 pp.

Penning de Vries, F.W.T. and van Keulen, H. (1982) La production actuelle et l'action de l'azote et du phosphore. In: Penning de Vries, F.W.T. and Djitèye, M.A. (eds) *La Productivité des Pâturages Sahéliens*. Agricultural Research Reports 918, Pudoc, Wageningen, pp. 196-226.

Penning de Vries, F.W.T., Brunsting, A.H.M. and van Laar, H.H. (1974) Products, requirements and efficiency of biosynthesis: a quantitative approach. *Journal of Theoretical Biology* 45, 339-77.

Penning de Vries, F.W.T., van Laar, H.H. and Chardon, M.C.M. (1983) Bioenergetics of growth of seeds, fruits, and storage organs. In: *Potential Productivity of Field Crops under Different Environments*. International Rice Research Institute, Los Banos, Philippines, pp. 37-59.

Penning de Vries, F.W.T., Jansen, D.M., ten Berge, H.F.M. and Bakema, A. (1989) *Simulation of Ecophysiological Processes of Growth in Several Annual Crops*. Simulation Monographs, Pudoc, Wageningen and International Rice Research Institute, Los Banos, 271 pp.

Penning de Vries, F.W.T., van Keulen, H. and Alagos, J.C. (1990)

Nitrogen redistribution and potential production in rice. In: Sinha, S.K., Sane, P.V., Bhargava, S.C. and Agrawal, P.K. (eds) *Proceedings of the International Congress of Plant Physiology*, 15-20 February 1988, New Delhi, India. Volume 1, InPrint Exclusives, S-402, Greater Kailash-II, New Delhi, pp. 513-20.

Pons, T.L. (1985) Growth rates and competitiveness to rice of some annual weed species. In: *Proceedings of the Seventh Asian-Pacific Weed Science Society Conference 1979. Biotrop Bulletin* 23, 13-21.

Porter, J.R. (1984) A model of canopy development in winter wheat. *Journal of Agricultural Science* (Cambr.) 102, 383-92.

Rabbinge, R. and Rossing, W.A.H. (1987) Decision models in pest management. *European Journal of Operational Research* 32, 327-32.

Rabbinge, R., Ward, S.A. and van Laar, H.H. (eds) (1989) *Simulation and Systems Management in Crop Protection*. Simulation Monographs 32, Pudoc, Wageningen, 420 pp.

Radosevich, S.R. and Holt, J.S. (1984) *Weed Ecology: Implications for Vegetation Management*. John Wiley and Sons, New York, 265 pp.

Ramakrishnan, P.S. and Kapoor, P. (1973) Photoperiodic requirements of seasonal populations of *Chenopodium album* (L.). *Journal of Ecology* 62, 67-73.

Ranganathan, R. (1993) *Analysis of Yield Advantage in Mixed Cropping*. PhD thesis, Department of Theoretical Production Ecology, Wageningen Agricultural University, 93 pp.

Rao, A.N. and Moody, K. (1987) Rice yield losses caused by transplanted *Echinochloa glabrescens* and possible control methods. In: *Proceedings of the 11th Asian-Pacific Weed Science Conference*, Taipei, Republic of China, pp. 203-10.

Rao, A.N. and Moody, K. (1992) Competition between *Echinochloa glabrescens* and rice (*Oryza sativa*). *Tropical Pest Management* 38(1), 25-9.

Rappoldt, C. and van Kraalingen, D.W.G. (1991) *FORTRAN Utility Library TTUTIL*. Simulation Report CABO-TT No. 20, Department of Theoretical Production Ecology, Wageningen Agricultural University, P.O. Box 430, 6700 AK Wageningen, 54 pp.

Rimmington, G.M. (1984) A model of the effect of interspecies competition for light on dry-matter production. *Australian Journal of Plant Physiology* 11, 277-86.

Rooney, J.M. (1991) Influence of growth form of *Avena fatua* L. on the growth and yield of *Triticum aestivum* L. *Annals of Applied Biology* 118, 411-6.

Ryel, R., Barnes, P.W., Beyschlag, W., Caldwell, M.M. and Flint, S.D. (1990) Plant competition for light analyzed with a multispecies canopy model. I. Model development and influence of enhanced

UV-B conditions on photosynthesis in mixed wheat and wild oat canopies. *Oecologia* 82, 304-10.

Samson, D.A. and Werk, K.S. (1986) Size-dependent effects in the analysis of reproductive effort in plants. *American Naturalist* 127, 667-80.

Seligman, N.G. and van Keulen, H. (1981) PAPRAN: A simulation model of annual pasture production limited by rainfall and nitrogen. In: Frissel, M.J. and van Veen, J.A. (eds) *Simulation of Nitrogen Behaviour of Soil-Plant Systems*. Pudoc, Wageningen, pp. 192-220.

Sheehy, J.E., Cobby, J.M. and Ryle, G.J.A. (1980) The use of a model to investigate the influence of some environmental factors on the growth of perennial ryegrass. *Annals of Botany* 46, 343-65.

Shinozaki, K. and Kira, T. (1956) Intraspecific competition among higher plants. VII. Logistic theory of the C-D effect. *Journal of Institute for Polytechnics, Osaka City University*, Series D 7, 35-72.

Silander Jr, J.A. and Pacala, S.W. (1985) Neighborhood predictors of plant performance. *Oecologia* 66, 256-63.

Sinclair, T.R. and Horie, T. (1989) Leaf nitrogen, photosynthesis, and crop radiation use efficiency: a review. *Crop Science* 29, 90-8.

Smith, R.J. (1968) Weed competition in rice. *Weed Science* 16, 252-5.

Smith, R.J. (1983) Weeds of major economic importance in rice and yield losses due to weed competition. In: *Weed Control in Rice*. International Rice Research Institute, Los Banos, Philippines, pp. 19-36.

Spitters, C.J.T. (1983a) An alternative approach to the analysis of mixed cropping experiments. 1. Estimation of competition effects. *Netherlands Journal of Agricultural Science* 31, 1-11.

Spitters, C.J.T. (1983b) An alternative approach to the analysis of mixed cropping experiments. 2. Marketable yields. *Netherlands Journal of Agricultural Science* 31, 143-55.

Spitters, C.J.T. (1984) A simple simulation model for crop-weed competition. *7th International Symposium on Weed Biology, Ecology and Systematics*. COLUMA-EWRS, Paris, pp. 355-66.

Spitters, C.J.T. (1986) Separating the diffuse and direct component of global radiation and its implications for modeling canopy photosynthesis. Part II. Calculations of canopy photosynthesis. *Agricultural and Forest Meteorology* 38, 231-42.

Spitters, C.J.T. (1989) Weeds: population dynamics, germination and competition. In: Rabbinge, R., Ward, S.A. and van Laar, H.H. (eds) *Simulation and Systems Management in Crop Protection*. Simulation Monographs, Pudoc, Wageningen, pp. 182-216.

Spitters, C.J.T. (1990) On the descriptive and mechanistic models for inter-plant competition, with particular reference to crop-weed interaction. In: Rabbinge, R., Goudriaan, J., van Keulen, H., Penning de Vries, F.W.T. and van Laar, H.H. (eds) *Theoretical Production Ecology: Reflections and Prospects*. Simulation Monographs 34, Pudoc, Wageningen, pp. 217-36.

Spitters, C.J.T. and Aerts, R. (1983) Simulation of competition for light and water in crop-weed associations. *Aspects of Applied Biology* 4, 467-84.

Spitters, C.J.T. and Kramer, Th. (1986) Differences between spring wheat cultivars in early growth. *Euphytica* 35, 273-92.

Spitters, C.J.T., Toussaint, H.A.J.M. and Goudriaan, J. (1986) Separating the diffuse and direct component of global radiation and its implications for modeling canopy photosynthesis. Part I. Components of incoming radiation. *Agricultural and Forest Meteorology* 38, 217-29.

Spitters, C.J.T., van Keulen, H. and van Kraalingen, D.W.G. (1989a) A simple and universal crop growth simulator: SUCROS87. In: Rabbinge R., Ward S.A. and van Laar H.H. (eds) *Simulation and Systems Management in Crop Protection*. Simulation Monographs, Pudoc, Wageningen, pp. 147-81.

Spitters, C.J.T., Kropff, M.J. and de Groot, W. (1989b) Competition between maize and *Echinochloa crus-galli* analysed by a hyperbolic regression model. *Annals of Applied Biology* 115, 541-51.

Stapper, M. and Arkin, G.F. (1980) *CORNF: a Dynamic Growth and Development Model for Maize (Zea mays L.)*. Texas Agricultural Experimental Station, Blackland Research Center, Temple, Texas, U.S.A., Program and Model Documentation 80-2, 91 pp.

Suehiro, K. and Ogawa, H. (1980) Competition between two annual herbs *Atriplex gmelini* C.A. Mey and *Chenopodium album* L. in mixed cultures irrigated with seawater of various concentrations. *Oecologia* 45, 167-77.

Takayanagi, S. and Kusanagi, T. (1991) Model for simulation of mechanisms of growth and light competition in *Digitaria ciliaris*-soybean mixed stands to devise an early method of diagnosis of weed damage. *Weed Research Japan* 36(4), 372-9.

Tanaka, A. (1983) Physiological aspects of productivity in field crops. In: *Potential Productivity of Field Crops under Different Environments*. International Rice Research Institute, Los Banos, Philippines, pp. 61-80.

Taylor, D.R. and Aarssen, L.W. (1989) On the density dependence of replacement-series competition experiments. *Journal of Ecology* 77, 975-88.

Terry, N. (1968) Development physiology of sugar beet. I. The influence of light and temperature on growth. *Journal of Experimental*

Botany 19, 795-811.

Thompson, B.K., Weiner, J. and Warwick, S.I. (1991) Size-dependent reproductive output in agricultural weeds. *Canadian Journal of Botany* 69, 442-6.

Thornley, J.H.M. (1972) A balanced quantitative model for root/shoot ratios in vegetative plants. *Annals of Botany* 36, 431-41.

Thornley, J.H.M. and Johnson, I.R. (1990) *Plant and Crop Modelling: A Mathematical Approach to Plant and Crop Physiology*. Clarendon Press, Oxford, 609 pp.

Thornton, P.K., Fawcett, R.H., Dent, J.B. and Perkins, T.J. (1990) Spatial weed distribution and economic thresholds for weed control. *Crop Protection* 9, 337-42.

Tilman, D. (1987) Secondary succession and the pattern of plant dominance along experimental nitrogen gradients. *Ecological Monographs* 57, 189-214.

Tilman, D. (1988) *Plant Strategies and the Dynamics and Structure of Plant Communities*. Monographs in Population Biology 20, Princeton University Press, Princeton, NJ, 360 pp.

Trenbath, B.R. (1976) Plant interactions in mixed crop communities. In: Stelly, M. (ed.) *Multiple Cropping*. American Society of Agronomy Special Publication No. 27, pp. 129-69.

Vandermeer, J.H. (1989) *The Ecology of Intercropping*. Cambridge University Press, Cambridge, 237 pp.

Versteeg, M.N. and van Keulen, H. (1986) Potential crop production prediction by some simple calculation methods, as compared with computer simulations. *Agricultural Systems* 19, 249-72.

Wahmhoff, W. and Heitefuss, R. (1988) Studies on the use of economic injury thresholds for weeds in winter barley. In: *Plant Research and Development; A Biannual Collection of Recent German Contributions*. Institut für Wissenschaftliche Zusammenarbeit, Tübingen, Germany, Volume 27, pp. 59-91.

Wall, P.C. (1983) The role of plant breeding in weed management in the advancing countries. In: *Improving Weed Management*. FAO Plant Production and Protection Paper 44, FAO, Rome, pp. 40-9.

Warwick, S.I. and Marriage, P.B. (1982) Geographical variation in populations of *Chenopodium album* resistant and susceptible to atrazine. II. Photoperiod and reciprocal transplant studies. *Canadian Journal of Botany* 60, 494-504.

Weaver, S.E. (1991) Size-dependent economic thresholds for three broadleaf weed species in soybeans. *Weed Technology* 5, 674-9.

Weaver, S.E., Smits, N. and Tan, C.S. (1987) Estimating yield losses of tomato (*Lycopersicon esculentum*) caused by nightshade (*Solanum* spp.) interference. *Weed Science* 35, 163-8.

Weaver, S.E., Kropff, M.J. and Groeneveld, R.M.W. (1992) Use of eco physiological models for crop-weed interference: The critical period of weed interference. *Weed Science* 40, 302-7.

Weir, A.H., Bragg, P.L., Porter, J.R. and Rayner, J.H. (1984) A winter crop simulation model without water or nutrient limitations. *Journal of Agricultural Science* 102, 371-82.

Wevers, J.D.A. (1991) Low dosage systems for controlling broad leaf and annual grass weeds in sugar beet. Mededelingen Faculteit Landbouwwetenschappen Rijksuniversiteit Gent 56, 611-5.

Wilkerson, G.G., Jones, J.W., Boote, K.J., Ingram, K.T. and Mishoe, J.W. (1983) Modeling soybean growth for crop management. *Transactions of the American Society of Agricultural Engineers* 26, 63-73.

Wilkerson, G.G., Jones, J.W., Coble, H.D. and Gunsolus, J.L. (1990) SOYWEED: A simulation model of soybean and common cocklebur growth and competition. *Agronomy Journal* 82, 1003-10.

Willey, R.W. (1979a) Intercropping - its importance and research needs. Part 1. Competition and yield advantages. *Field Crop Abstracts* 32(1), 1-10.

Willey, R.W. (1979b) Intercropping - its importance and research needs. Part 2. Agronomy and research approaches. *Field Crop Abstracts* 32(2), 73-85.

Willigen, P. de (1991) Nitrogen turnover in the soil-crop system; comparison of fourteen simulation models. *Fertilizer Research* 27, 141-9.

Willigen, P. de and Neeteson, J.J. (1985) Comparison of six simulation models for the nitrogen cycle in the soil. *Fertilizer Research* 8, 157-71.

Wit, C.T. de (1960) *On Competition*. Agricultural Research Reports 66.8, Pudoc, Wageningen, 82 pp.

Wit, C.T. de (1965) *Photosynthesis of Leaf Canopies*. Agricultural Research Reports 633, Pudoc, Wageningen, 57 pp.

Wit, C.T. de (1982) Simulation of living systems. In: Penning de Vries, F.W.T. and van Laar, H.H. (eds) *Simulation of Plant Growth and Crop Production*. Simulation Monographs, Pudoc, Wageningen, pp. 3-8.

Wit, C.T. de and Goudriaan, J. (1978) *Simulation of Ecological Processes*. Simulation Monographs, Pudoc, Wageningen, 175 pp.

Wit, C.T. de and Penning de Vries, F.W.T. (1982) L'analyse des systèmes de production primaire. In: Penning de Vries, F.W.T. and Djitèye, M.A. (eds) *La Productivité des Pâturages Sahéliens*. Agricultural Research Reports 918, Pudoc, Wageningen, pp. 20-7.

Wit, C.T. de and Penning de Vries, F.W.T. (1985) Predictive models in agricultural production. *Philosophical Transactions of the Royal Society* London, B 310, 309-15.

Wit, C.T. de, Goudriaan, J., van Laar, H.H., Penning de Vries, F.W.T., Rabbinge, R., van Keulen, H., Sibma, L. and de Jonge, C. (1978) *Simulation of Assimilation, Respiration and Transpiration of Crops.* Simulation Monographs, Pudoc, Wageningen, 141 pp.

Wolf, J., de Wit, C.T., Janssen, B.H. and Lathwell, D.J. (1987) Modeling long-term crop response to fertilizer phosphorus. I. The model. *Agronomy Journal* 79, 445-51.

Wong, S.C., Cowan, I.R. and Farquhar, G.D. (1985) Leaf conductance in relation to rate of CO_2 assimilation. *Plant Physiology* 78, 821-34.

Wright, A.J. (1981) The analysis of yield-density relationships in binary mixtures using inverse polynomals. *Journal of Agricultural Science* (Cambr.) 96, 561-7.

Wymore, L.A. and Watson, A.K. (1989) Interaction between a velvetleaf isolate of *Colletotrichum coccodes* and thidiazuron for velvetleaf (*Abutilon theophrasti*) control in the field. *Weed Science* 37, 478-83.

Yoda, K., Kira, T., Ogawa, H. and Hozumi, K. (1963) Self-thinning in overcrowded pure stands under cultivated and natural conditions (Intraspecific competition among higher plants. XI). *Journal of Biology, Osaka City University* 14, 107-29.

Yoshida, S. (1981) *Fundamentals of Rice Crop Science.* International Rice Research Institute, Los Banos, Philippines, 269 pp.

Zimdahl, R.L. (1980) *Weed-Crop Competition - a Review.* International Plant Protection Center, Oregon State University, Corvallis.

Zimdahl, R.L. (1988) The concept and application of the critical weed-free period. In: Altieri, M.A. and Liebman, M. (eds) *Weed Management in Agroecosystems: Ecological Approaches.* CRC Press Inc., Boca Raton, Florida, pp. 145-55.

Appendix One

The FORTRAN Simulation Environment (FSE)

D.W.G. van Kraalingen

Introduction

The model INTERCOM presented in this book is programmed in FORTRAN 77, using a crop growth simulation shell called FSE (FORTRAN Simulation Environment; van Kraalingen, 1991). FSE was developed to provide crop modellers with a programming environment in FORTRAN 77 (see Appendix 2, Listing A2.1). The simulation environment consists of a main model that contains the control structure for rerun facilities, reading of weather data and the dynamic loop (integration, rate calculation and time update), a framework for the major process related routines and a collection of utility routines that perform specific tasks such as reading of parameter values from data files and for generating model output.

The utility routines used are part of the FORTRAN 77 library TTUTIL (Rappoldt and van Kraalingen, 1991). The reports by van Kraalingen (1991) and Rappoldt and van Kraalingen (1991) can be obtained from the Department of Theoretical Production Ecology, Wageningen Agricultural University, P. O. Box 430, 6700 AK Wageningen, The Netherlands.

Integration and time loop

The integration method used in the FSE program is the Euler or rectangular integration method. The order in which calculations are executed and how reruns are implemented is shown in Fig. A1.1. At the point where output is generated, values of state variables and rate variables refer to the same time. In the design of FSE, state and rate calculations are implemented in separate sections in the major subroutines for plant and soil processes. The main program controls which section is activated through the concept of *task-controlled exe-*

187

cution. This is illustrated in Fig. A1.2. The program lines of the plant and soil water subprocesses are separated into rate and state sections and only one of these tasks is executed during a single call from the main program. Four different tasks are distinguished: *initialization* (ITASK=1), *integration* (ITASK=3), *rate calculation* (ITASK=2) and *terminal calculation* (ITASK=4).

After each time step a decision is made if another time step is required or if the simulation should proceed to the terminal section (Fig. A1.1). One of the criteria to stop the simulation is that the predefined finish time (FINTIM) has been exceeded. In crop growth models, however, simulation has to be terminated when the crop is mature or if some other criterion has been met. It is thus necessary that the simulation loop can be terminated from within each of the submodels (e.g. subroutine PLANTC, Listing A2.1). This is implemented using a variable called TERMNL of the type LOGICAL, which indicates if the simulation loop should be terminated. The simulation loop continues as long as TERMNL=.FALSE. and the criterion is programmed as an emulated DO-WHILE loop.

The subroutine TIMER (TTUTIL library) controls the time variables in the model. The basic actions of the subroutine TIMER are: (*i*) (ITASK=1) check the values of FINTIM, DELT (time step of integration), TIME, etc. and copy these to variables local to the subroutine,

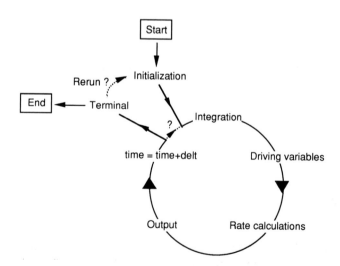

Fig. A1.1. The order in which calculations are executed when simulating continuous systems using the Euler integration, illustrating where to enter and to leave the circle and how reruns are implemented.

switch on the output flag at the start of the simulation when TIME is a multiple of PRDEL (time interval for printed output) and when the simulation terminates, (*ii*) (ITASK=2) check whether the local time variables have the same value as the global time variables, add DELT to TIME, calculate the day number (DAY), flag if TIME is a multiple of PRDEL using the variable OUTPUT, flag if TIME has exceeded FINTIM using the variable TERMNL.

The TIMER routine which is used in the FSE program has some extra features: the year of simulation is automatically incremented and leap years are also recognized (DAY runs until 366), and the day number is available as an integer and as a real value (IDAY and DAY, respectively).

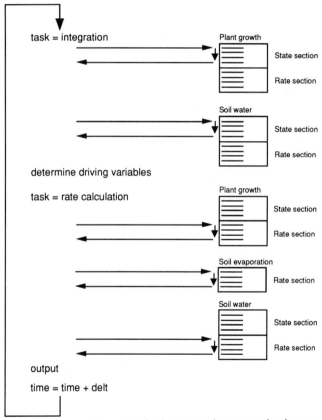

Fig. A1.2. General structure for incorporating several subprocesses into a single simulation model. The plant growth and soil water rectangles represent *one* subprocess description containing integration and rate calculations but called from different places in the main model with different task parameters.

Initialization of the states and parameters from external data files

All state variables in the model have to be initialized in the process subroutines (like subroutine PLANTC, Listing A2.1). The rate variables do not have to be initialized, because the model starts with the rate calculations after the initialization. Integration is only performed if previously a rate calculation has been carried out.

The input data files: reading data

Most of the parameters and initial values of the state variables of the various subprocesses are read from data files. The data file TIMER.DAT (Listing A1.1), the files with species characteristics (SUGARB.DAT, CHENO.DAT; Listings A2.12 and A2.13, respectively) and SOIL.DAT (Listing A1.2) have identical formats, and each variable in them may appear only once (the weather data files will be discussed separately). The values for the initial state variables and parameters are read from the data files using a set of TTUTIL subroutines whose names all begin with RD (e.g. RDSREA means 'read a single real value', see also Table A2.1). With these routines the user can request the value by supplying the name of the requested variable (after having defined which data file to use). The statement:

```
CALL RDSREA ('WLVGI', WLVGI)
```

requests the subroutine RDSREA to extract the value of WLVGI from the data file and assigns it to the variable WLVGI. The data file is selected by the following statement:

```
CALL RDINIT (IUNITP, IUNITO, FILEP)
```

which calls the routine that
- opens the file with variable name FILEP using unit = IUNITP and IUNITP+1 (FILEP is a character string that has been assigned to the string PLANT.DAT (here, e.g. CHENO.DAT) in the calling program), then
- analyses the data file,
- creates a temporary file from the data file using unit=IUNITP,
- closes the data file (leaving IUNITP used !!), and
- sends all error messages that have been created to a log file (with unit=IUNITO).

After this call, the plant subroutine can acquire the numerical values (including arrays) or character strings through the different RD routines, RDSREA (read single real), RDSINT (read single integer), RDAREA (read array of reals), and RDSCHA (read single character string (in TIMER.DAT)). The CLOSE statement deletes the temporary file that is created by the RD routines. A corresponding data file PLANT.DAT could be as follows:

```
WLVGI   = 0.; AMAXM = 40.
AMDVST  = 0.,1., 1300., 2200.,0.
KDF     = 0.69
```

The following syntax rules apply to these input files:
- The file consists of names and numerical values of variables, separated by an '=' sign, e.g. AMAX = 40.;
- The name of a variable may not exceed six characters;
- For array variables, more than one numerical value may follow the equal sign, separated by commas;
- Identical numerical array values can be given as n*<numerical value> (e.g. 10*5.4);
- Variables may appear in the file in any order;
- Comment lines start with '*' in the first column, or '!' in any column (the rest of the line is ignored);
- Continuation character is ',' on preceding line (applies to arrays only);
- Names of variables and numerical values can be given on the same line if separated by a single semicolon ';' ;
- Only the first 80 characters of each record of the data file are read;
- No tabs or other control and extended ASCII characters are allowed in the file.

The TIMER.DAT file

The input file TIMER.DAT (Listing A1.1) specifies the value of the time variables such as time step of integration, time between different outputs to file, etc.; the directory in which the weather data are stored, the country code, station number and year; and some miscellaneous control variables. If many weather data files are used, it may be convenient to store these data in a separate directory. By assigning a directory name to the variable WTRDIR, the weather system is directed to read weather data from that directory, e.g.:

```
WTRDIR = 'C:\WEATHER\'
```

Appendix 1

The country code:

```
CNTR = 'NL'
```

For a list of available weather data files, their corresponding country codes and station numbers, see van Kraalingen *et al.* (1990), e.g.

NL   <- country code for the Netherlands
PHIL   <- country code for the Philippines

These two character strings are read through the RDSCHA routine (TTUTIL library) in the MAIN program.

Listing A1.1. Example of the contents of a TIMER.DAT file. For explanation of abbreviations see Appendix 3.

```
****************************************************************
*                                                              *
* Defining the simulation run                                  *
*                                                              *
****************************************************************
*
WTRDIR = 'C:\WEATHER\'
CNTR   = 'NL'
ISTN   = 1        ! Station number of weather data
IYEAR  = 1985     ! Year of weather data
*
* Time variables and output file options
DAYB   = 121.     ! Start day of simulation
FINTIM = 148.     ! Finish time of simulation
PRDEL  = 5.       ! Time between consecutive outputs to file
DELT   = 1.       ! Time step of integration
ITABLE = 4        ! Format of output file
                  ! (0 = no output table, 4 = normal table,
                  !  5 = Tab-delimited (for Excel), 6=TTPLOT format)
IDTMP  = 0        ! Switch variable what should be done with the
                  ! temporary output files (0 = do not delete,
                  ! 1 = delete)
IRUNLA = 0        ! 1=LAI measured, 0=LAI simulated
HARDAY = 269.
                  ! List of harvest data for which output is wanted
                  ! 0. = no harvest data
                  ! example: HARDAY=100.,140.,145.,183.
*
* Definition competing species:
IPSPEC = 1    , 2        ! 1=SUGAR BEET, 2=CHENOPODIUM
IDAYEM = 129  , 141      ! Dates of emergence
NPL    = 11.11, 5.5      ! Plant densities
```

The variables ISTN and IYEAR refer to the weather data and indicate the station number and year from a country. For example, when the country code is NL (the Netherlands), ISTN=1 and IYEAR=1984, the weather data from Wageningen 1984 will be used by the model.

During execution, the weather system will try to open a file by the name of NL1.984 on the given directory given (WTRDIR).

The variables DAYB, FINTIM, PRDEL and DELT represent the time parameters of the model. DAYB is the start day of the whole program; its value should be between 1 and 365. FINTIM is the finish time of the simulation, counted from the start of simulation. For example, when DAYB=93., and FINTIM=10., the simulation will continue until DAY=103. The variable PRDEL indicates the time between consecutive outputs to the output file (RESULTS.OUT). For example, when PRDEL=5., output is given each time that TIME is a multiple of PRDEL (TIME=5.,10.,15. etc.). Irrespective of the value of PRDEL, output is always given at the start of the simulation (TIME=0) and when the simulation is terminated (either FINTIM=TIME or some other finish criterion). By giving PRDEL a high value (e.g. 1000), intermediate outputs are suppressed. The value of DELT, the time step of integration, is one day. This value cannot be changed, because of the procedures used in the CO_2 assimilation subroutines, that calculate daily rates using the Gaussian integration method.

The variable ITABLE defines if an output table is required (no output table: ITABLE=0) and if so, what the format should be. A multiple column table (ITABLE=4) is sufficient for normal printing and viewing. Using ITABLE=5, a tab-delimited multiple column table, which is easily imported in spreadsheet programs such as EXCEL, is generated. A two-column format is generated using ITABLE=6.

The variable IDTMP defines whether the file with temporary output data (RES.BIN) should be deleted at termination of the simulation (IDTMP=0, do not delete, IDTMP=1, delete). This file is created during the dynamic phase of the simulation and is used during the terminal phase of the simulation to generate the output file RESULTS.OUT.

The variable HARDAY can be used to force output at day numbers for which harvest data from the field are available. In many cases these harvest data will not coincide with output intervals in the model unless PRDEL is set to unity (which may cause large output files to be generated). A maximum of 20 day numbers can be defined here. A single value of zero indicates that no forced output is required, e.g.:

```
HARDAY = 0.                          <- No forced output
HARDAY = 11.,27.,52.     <- Output is forced on days 11, 27 and 52
```

The variable IRUNLA defines whether the leaf area index (*LAI*) used in the model is a forcing function based on field measurements (IRUNLA=1) or whether the *LAI* is simulated (IRUNLA=0).

The variable `IPSPEC` controls how many and which plant species compete in the model. A list of numbers indicating crop species can be typed after the equal sign. For instance if sugar beet is defined as '1' and *C. album* as '2', the examples are:

```
IPSPEC = 1              <- Sugar beet monoculture simulation
IPSPEC = 2              <- C. album monoculture simulation
IPSPEC = 1,2    <- Simulation of sugar beet /C. album competition
```

The numbers refer to a species data file as defined in the `PLANTC` routine. Related to the `IPSPEC` variable are the variables `IDAYEM` and `NPL`, because these indicate the emergence dates and number of plants per m^2 for each of the species defined with `IPSPEC`. The number of values supplied here should, therefore, be equal to the number of `IPSPEC`. For example:

```
IPSPEC = 1, 2    <- Simulation of sugar beet / C. album competition
IDAYEM = 124, 137       <- -Sugar beet emergence at day 124,
                          and C. album emergence at day 137
NPL    = 11., 9.7     <- Number of sugar beet plants : 11 m-2
                        and number of C. album plants: 9.7 m-2
```

The RERUNS.DAT file

If the file `RERUNS.DAT` is absent or empty, the model will execute a single run, using the data from the standard data files. An example is given in Listing A1.2. By creating a rerun file, the model will execute additional runs with different parameters and/or initial values for the state variables. Therefore, the total number of runs made by the model is always one more than the number of rerun sets. Names of variables originating from different data files can be redefined in the same rerun file. The format of the rerun files is identical to that of the other data files, except that the names of variables may appear in the file more than once. Arrays can also be redefined in a rerun file. The order and number of the variables should be the same in each set. A new set starts when the first variable is repeated. This is shown in the following example, where the variables `IDAYEM` and `NPL` from file `TIMER.DAT` are redefined:

```
IDAYEM =  90, 110;  NPL = 11., 5.      <- 1st rerun set
IDAYEM = 110,  90;  NPL = 11., 5.      <- 2nd rerun set
```

Important:

1. Each variable of which the value is changed somewhere in the re-run file should be assigned a value in each set, even if that value is identical to the value in the previous set;
2. Variables that are indexed for more species (e.g. AMAX(1)) cannot be used in the RERUNS file.

Listing A1.2. Example of the contents of a RERUNS.DAT file.

```
*********************************************************************
* RERUNS.DAT file                                                   *
*                                                                   *
* When a value for an input variable is changed in a rerun,         *
* that value must be changed accordingly in each set, even          *
* if that value is identical in all sets.                           *
*                                                                   *
* Lines with an asterisk (*) in the first column are comment lines  *
* Input variables can be separated by a semi-column (;)             *
*                                                                   *
*********************************************************************

IPSPEC = 2                   ! 1st run,  2 = C. album
IDAYEM = 141                 !           Date of emergence
NPL    = 11.11               !           Plant density

IPSPEC = 1, 2                ! 2nd run,  1 = sugar beet, 2 = C.album
IDAYEM = 129, 141            !           Date of emergence
NPL    = 11.11, 5.5          !           Plant densities
*********************************************************************
* If you want to change this RERUNS.DAT file:                       *
* Replace the above series by other variables and/or values         *
* NOTE: only variables from TIMER.DAT can be used for reruns!        *
*********************************************************************
```

The PLANT.DAT and SOIL.DAT files

With the FSE competition model, plant data files and a soil data file are supplied containing parameters and initial state values: e.g.

Listing A1.3. Example of the contents of a SOIL.DAT file.

```
*********************************************************************
* Soil characteristics                                              *
*********************************************************************
RIRRIT = 1.,0., 365.,0.
RCAPT  = 1.,0., 180.,0., 181.,1., 195.,1., 196.,0., 365.,0.
THCKT  = 0.02
SMTI   = 5.
SMRTZI = 182.
TCS    = 1.
RTDMAX = 0.6
VSMWP  = 89.    ! Wilting point in liters per m3
VSMFC  = 303.   ! Field capacity in liters per m3
*********************************************************************
```

SUGARB.DAT containing sugar beet data, CHENO.DAT containing *C. album* data and SOIL.DAT containing irrigation and soil data (Listing A1.3). These parameters and initial state variables are discussed at length in this book and will not be discussed further here. The general syntax rules for data files as discussed above apply to these files as well. The plant input files are given in Appendix 2.

Weather data

The weather data used in the model are read from external files. The weather data file definition, however, is different from those for the RD routines. The weather data system used has been developed jointly by the Centre for Agrobiological Research and the Department of Theoretical Production Ecology of the Wageningen Agricultural University. It is especially designed for use in crop growth simulation models and has been documented in a separate report (van Kraalingen *et al.*, 1991; available on request).

The weather data system basically consists of two parts: the weather data files and a reading program to retrieve data from those files. A single data file can contain, at the most, the daily weather data from one meteorological station for one particular year. The country name (abbreviated), station number and year to which the data refer are reflected in the name of the data file (Listing A1.4).

The reading program consists of a set of subroutines and functions, only two of which are intended to be called by the main program (STINFO and WEATHR, see Listing A2.1). The others are internal to the reading program.

A call to the first subroutine (STINFO) defines the country (CNTR), station code (ISTN), year number (IYEAR) and the name of the directory containing the weather data (WTRDIR), this information is first read from the data file TIMER.DAT. A control parameter (IFLAG) is also supplied to indicate where possible messages of the system should be directed (screen and/or log file), and a name must be given to the log file if that name should differ from the default name WEATHER.LOG. The subroutine STINFO returns the location parameters (longitude LONG, latitude LAT, and altitude ELEV) of the selected meteorological station and, if the radiation is calculated from sunshine hours by the weather system, two coefficients of the Ångström formula (*a* and *b*) pertaining to the selected station. The value of a status variable (ISTAT) indicates a possible error or warning (e.g. the data file requested does not exist). The location parameters can later be used to calculate day length (from the latitude) or average air pressure (from the altitude).

Listing A1.4. Example of the contents of a WEATHER file (NL1.985).

```
*-------------------------------------------------------------------*
* Station name: Wageningen (Haarweg), Netherlands
* Year: 1985
* Author: Peter Uithol                         -99.000: NIL VALUE
* Source: Natuur- en Weerkunde via Nel van Keulen
* Longitude: 5 40 E, latitude: 51 58 N, altitude: 7 m.
*
* Column  Daily value
* 1        station number
* 2        year
* 3        day
* 4        irradiation                 (kJ m-2 d-1)
* 5        minimum temperature         (degrees Celsius)
* 6        maximum temperature         (degrees Celsius)
* 7        early morning vapour pressure (kPa)
* 8        mean wind speed (height: 2 m) (m s-1)
* 9        precipitation               (mm d-1)
*-------------------------------------------------------------------*
    5.67  51.97      7. 0.00 0.00
   1 1985   1   660.    0.2    5.7   0.670   5.4   6.8
   1 1985   2  2200.   -2.9    0.7   0.490   2.2   0.1
   1 1985   3  2280.   -5.5    0.2   0.520   1.6   0.0
   1 1985   4  3310.  -18.4   -5.2   0.230   0.7   0.4
   1 1985   5  4220.  -19.5   -7.8   0.210   0.6   0.0
   1 1985   6  1940.  -13.8   -6.2   0.310   3.9   1.7
   1 1985   7  4040.  -21.4   -9.7   0.200   3.3   0.0
   1 1985   8  2710.  -21.3   -6.2   0.240   2.2   0.0
   1 1985   9   930.   -7.8   -3.8   0.380   4.1   0.1
   1 1985  10  1540.  -11.6   -3.6   0.390   2.1   0.6
   1 1985  11  2380.   -5.3   -3.5   0.430   2.0   0.0
   1 1985  12  2310.   -7.9    1.7   0.470   3.3   1.6
   1 1985  13  3800.  -13.9   -5.5   0.270   2.6   0.0
   1 1985  14  2880.  -12.9   -3.9   0.310   2.7   0.1
   1 1985  15  2750.  -11.2   -9.3   0.240   2.8   0.0
   1 1985  16  1540.  -15.0   -6.6   0.270   1.0   0.0
   1 1985  17  2230.  -10.8   -5.1   0.310   1.2   0.3
   1 1985  18  3440.  -13.7   -6.1   0.260   1.9   0.0
   1 1985  19  1610.  -11.7   -5.3   0.340   1.7   0.0
   1 1985  20  1340.   -8.2   -3.0   0.370   2.4   0.0
   1 1985  21  1930.   -3.6    5.1   0.650   4.3   2.2


   1 1985 362   940.   -6.2   -3.0   0.430   1.1   0.0
   1 1985 363  2000.   -3.7   -1.2   0.460   1.7   0.0
   1 1985 364   900.   -5.3   -2.4   0.450   2.9   0.0
   1 1985 365  3410.   -6.2   -3.0   0.660   5.4   0.0
*-------------------------------------------------------------------*
```

After this initialization procedure, weather data for specific days can be obtained by calls to the second subroutine (WEATHR) with day number starting from January 1st as 1, as input parameter. The output of WEATHR consists of six weather variables for that day and the value of the status variable ISTAT indicating a possible error or warning (e.g. missing data, data obtained by interpolation, requested day is out of range, etc.). The six weather variables are daily incoming total global radiation (DRAD), minimum and maximum air temperature

(TMN and TMX), vapour pressure (VAPOUR), wind speed (WIND) and rainfall (RAIN).

The subroutine STINFO can be called again at any time during program execution to change any of its input parameters. A call to STINFO with identical input parameters is also permitted (in fact this is done regularly in the FSE main program). Similarly, the subroutine WEATHR can be called repeatedly with any day number between 1 and 365 (or 366 in the case of a leap year).

Implementing reruns

The FSE program includes a rerun facility. The general idea behind the rerun facility is that the data files remain identical and that the changes in data are specified in a separate file called RERUNS.DAT which may contain the names and values of variables from any of the 'standard' data files that are read by the program. Thus, the file RERUNS.DAT may contain parameters from SOIL, PLANT and TIMER data files (if they are not indexed!). Thus, in the INTERCOM model only data from SOIL.DAT and TIMER.DAT can be used. In the first run, the values from the standard data files will be used. In subsequent runs those values are then automatically replaced by the values from the rerun file. Execution will continue until all the rerun sets from RERUNS.DAT have been used. The output of the different runs is merged in one output file. In the INTERCOM model basically only variables from the TIMER.DAT file can be used.

It is shown in Fig. A1.1 that the control structure for the reruns is programmed as a loop around the actual model.

The call to RDSETS determines the presence of the RERUNS.DAT file and then analyses the data file if there is one. The return variable INSETS contains the number of rerun sets present in the rerun file; its value is zero if the rerun file is absent or empty. The subsequent DO-loop runs INSETS+1 times, because there is always one more than the number of sets in the rerun file (one run with standard data files + INSETS reruns). The value of the DO-loop counter (I1, the set number) is then used in the call to RDFROM to select a parameter set for the simulation. For I1 is zero, the standard data files will be used by the RD routines, for I1 larger than zero, the RD routines will automatically replace values with values from the rerun file.

Results from the reruns are written to the output file after analysis of the rerun file and after each replacement. Before a rerun is started, the model checks if all the variables of the preceding set were used. If this is not the case, it is assumed that there is a typing error in the data files and the simulation is halted.

Output of simulation results

Output is organized from each major subroutine (like the subroutines PLANTC and WBAL), separately. This avoids large argument lists to communicate output variables to the main program and limits the number of changes in the main program when, for instance, another plant routine with different output variables is used.

All subroutines write their output to the same output file (of which the name is defined in the FSE main program). By using a set of special routines (the OUT routines), output can be written in the form of tables. It is also possible to add print plots of selected variables to that output file (for details see van Kraalingen, 1991). The use of the OUT routines considerably simplifies the generation of output files. The available routines are OUTDAT for output of single variables and OUTARR for arrays.

The OUTDAT routine also has a task parameter as input (the first argument in the call statement), similar to the subprocess descriptions. The first call (with ITASK=1) to OUTDAT (CALL OUTDAT(1,20,'X',0.)) specifies that X will be the independent variable and that unit 20 and 21 can be used for the output file. Subsequent calls with ITASK=2 (CALL OUTDAT(2,0,'X',X)) instruct OUTDAT to store the incoming names and numerical values in a temporary file (with the units from the ITASK=1 call). The number of combinations of name and value that can be stored depends solely on free disk space and not on RAM memory. The first call to OUTDAT below the DO-loop (with ITASK=4, CALL OUTDAT(4, 0,'X',0.)) instructs the routine to create an output table using the information stored in the temporary file. Different output formats may be chosen, dependent on the value of the task variable. Tab-delimited format (e.g. for the EXCEL spreadsheet) can be generated by defining ITABLE in the TIMER.DAT file ITABLE=5, two-column format with ITASK=6. With any of these ITASK values, the string between quotes is written above the output.

The OUTARR call is actually an 'interface' call to OUTDAT. What the routine does internally is that it generates names (like A(1) and A(2)) and calls OUTDAT repeatedly for each of these name-value combinations. The range of array subscripts that should be generated by OUTARR is specified by the third and the fourth (last) subroutine arguments. This procedure can be repeated several times. The final call to OUTDAT (with 99) deletes the temporary file (CALL OUTDAT(99,0,' ',0.)).

Operation of the model

The model does not require interactive input during execution. The runs have been specified completely in the data files. During execution, the model will display run number, year number and day number on the screen. During execution, errors and warnings may occur from the weather system and/or from the other modules of the model. They generally consist of one line of text. If simulation is terminated by an error during the dynamic section of the run, the outputs generated before the error in that particular run occurred are written to the temporary file but are not yet written to the output file (RESULTS.OUT) until the terminal section of the model. Data can be recovered from the temporary file, using the OUTREC program (OUTput RECovery, see next Section).

Errors and warnings from the FSE program

Several checks are performed by the model. All errors terminate the model execution and a message to that effect is displayed on the screen. In some cases the error is also written to the output file (RESULTS.OUT). Warnings are displayed on the screen and are sometimes also written to the output file.

The weather system can also generate errors and warnings. Unlike errors from other sections of the model, the weather system itself never terminates the execution of the model. It is the FSE MAIN program that subsequently terminates the simulation run. Errors from the weather system are written to the screen and the log file WEATHER.LOG, warnings are written to the log file only.

If a run is terminated by some error from the model, the output file RESULTS.OUT will not contain the results of that specific run. But the results up until the error occurred are written to the temporary file RES.BIN. This file can be converted into an output table by running the output recovery program OUTREC. This program requests an integer number from the user. A standard output table is generated by a '4' (the default), '5' generates a tab-delimited table (meant to be imported in EXCEL), '6' generates an output of only two columns at a time. The output table will be written to the file OUTREC.OUT so that any existing RESULTS.OUT file is not deleted.

Appendix Two

Program Structure of the Model INTERCOM

M.J. Kropff, D.W.G. van Kraalingen and H.H. van Laar

Introduction

The model INTERCOM is implemented in the FORTRAN Simulation Environment (FSE, Appendix 1) to obtain a clear model structure and standardized input and output formats. Fig. A2.1 gives an overview of the structure of the computer program of the model. The model INTERCOM consists of the MAIN program which calls the FSE control subroutines (TIMER, STINFO and WEATHR, together with a group of utility subroutines) and the subroutines PLANTC, WBAL and DEVAP. The subroutine WBAL simulates the water balance of the soil, DEVAP simulates soil evaporation and PLANTC simulates growth and transpiration of the competing species.

During execution of the program, the program is initialized (ITASK=1, see Appendix 1). In the subroutine TIMER, initial values are assigned to the variable TIME and time related variables, like PRDEL (time interval of printed output) and FINTIM (finish time of the simulation run) through the settings in the file TIMER.DAT. Values of weather data and the coordinates of the location are read by the subroutines WEATHR and STINFO, respectively. Subsequently, the variables in the subroutines WBAL and PLANTC are initialized.

In the dynamic part of the program the rates are integrated first (ITASK=3) in the subroutines WBAL and PLANTC and subsequently the values of the rate variables (ITASK=2) are calculated. Therefore, the weather data are read and the subroutines PLANTC, DEVAP and WBAL are called. The time is updated by calling the subroutine TIMER and then the daily integration and rate calculation loop is continued until the finish time is reached.

In the terminal section (ITASK=4) the output is generated according to the assignments given in the TIMER.DAT file.

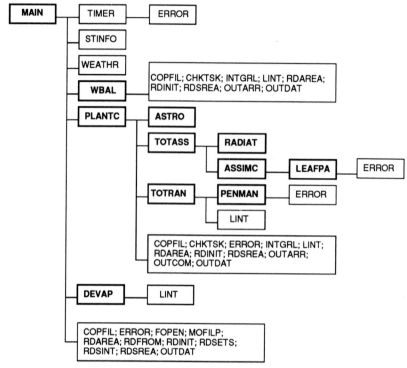

Fig. A2.1. Structure of the subroutines in the model INTERCOM. The bold boxes are called from the FSESUB library and the others from the TTUTIL library. In Table A2.1 the functions of the subroutines are described.

Table A2.1. Subroutines and functions called from the TTUTIL library in the MAIN program, and subroutines called from the FSESUB library in the subroutine PLANTC.

MAIN program:

CHKTSK	Checks the new task and previous task in the subroutines where ITASK defines which section is active (TTUTIL).
COPFIL	Copies contents of file with unit number IIN and name FILE into an output file with unit number IOUT (the output file should already be opened), only the input file is closed after the contents have been copied (TTUTIL).
ERROR	Writes error messages to the screen (TTUTIL).
DEVAP	Computes daily soil evaporation (FSESUB).
FOPEN	Opens a file after inquiry about the existence of the file (TTUTIL).
INTGRL	Function that executes the Euler integration method (TTUTIL).
LINT	Linear interpolation function (TTUTIL).

MOFILP	Moves the file pointer across comment lines of data files (beginning with '*') during reading of the file, and puts the file pointer at the first non-comment record (TTUTIL).
OUTARR	Stores an array which is written to the output file generated by OUTDAT (TTUTIL).
OUTCOM	Stores a text string which is written to the output file generated by OUTDAT (TTUTIL).
OUTDAT	Generates output file of the simulation model (TTUTIL).
RDAINT	Reads an array of integers from a data file (TTUTIL).
RDAREA	Reads an array of reals from a data file (TTUTIL).
RDINIT	Initializes subroutine RDDATA (not listed here) where values are assigned to variables in a data file (TTUTIL).
RDSCHA	Reads a single character string from a data file (TTUTIL).
RDSETS	Initializes subroutine RDDATA for reading data from a rerun data file (TTUTIL).
RDSINT	Reads a single integer value from a data file (TTUTIL).
RDSREA	Reads a single real value from a data file (TTUTIL).
STINFO	Sets weather system parameters, e.g. location, latitude weather station (TTUTIL).
TIMER	Updates the variable TIME and related variables (TTUTIL).
WBAL	Simple water balance of the soil, in which the actual volumetric soil moisture content of the top layer and of the root layer is calculated (FSESUB).
WEATHR	Reads weather data (TTUTIL).

Subroutine PLANTC:

ASTRO	Computes day length and photoperiodic day length from day number and latitude (FSESUB).
ASSIMC	Computes instantaneous gross canopy CO_2 assimilation and amount of absorbed radiation for a mixed canopy (FSESUB).
LEAFPA	Computes leaf area index above the defined height and leaf area density assuming a parabolic leaf area distribution (FSESUB).
LEAFRE	Computes leaf area index above the defined height and leaf area density assuming a rectangular leaf area distribution (FSESUB).
PENMAN	Computes potential evapo(transpi)ration (FSESUB).
RADIAT	Computes diffuse and direct incoming flux of photosynthetically active radiation from daily average global radiation (FSESUB).
TOTASS	Computes daily total gross canopy CO_2 assimilation (FSESUB).
TOTRAN	Computes the potential and actual transpiration rate of the species in a mixed canopy, the total actual transpiration rate of the canopy and the factor accounting for the water stress on the rate of dry matter increase (FSESUB).

Note: The subroutines WBAL and DEVAP are also called from FSESUB!

Data files needed to operate the model INTERCOM

To run the model INTERCOM several external data files are needed (Fig. A2.2). The variables and functions describing the plant species characteristics are read from the plant data files, e.g. for *C. album*

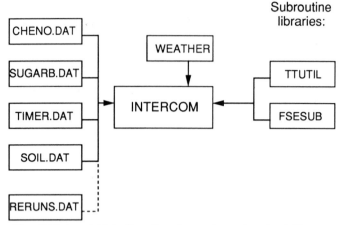

Fig. A2.2. Input (data files, libraries and weather data) files, necessary to execute the model INTERCOM.

(CHENO.DAT, Listing A2.13) and for sugar beet (SUGARB.DAT, Listing A2.12).

In the TIMER.DAT file (Appendix 1, Listing A1.1) the simulation run is defined. In this file the directory where the weather data are stored and the country code are indicated, time variables and output options are specified. A special variable indicates whether the leaf area index is simulated (IRUNLA=0) or that measured values are input in the model (IRUNLA=1). The variable HARDAY facilitates the possibility to compare simulation results with observed data on harvest dates. Also the definition of the competing plant species is given (in mixture IPSPEC=1,2 or IPSPEC=1,1; in monoculture IPSPEC=1 or IPSPEC=2). The dates of emergence and the plant densities have to be specified as well (e.g. monoculture IDAYEM=100 and NPL=10.; mixture IDAYEM=100,100 and NPL=10.,50.).

In the file SOIL.DAT the soil characteristics are defined.

Use of the RERUNS.DAT file is optional. By creating this file, the model can execute additional runs for different values of the variables only from the TIMER.DAT file (Note: changing the value of a variable for one of the plant data files in the RERUNS.DAT will result in the same assignment for all plant data sets!).

All utility subroutines are stores in a library called TTUTIL and all program subroutines, except the MAIN program and PLANTC (being the model INTERCOM) are stored in a library called FSESUB.

Listings of the model INTERCOM, subroutines and plant data

Listing A2.1. Listing of the main program and subroutine PLANTC of the model INTERCOM. For explanation of abbreviations see Appendix 3.

```
*=====================================================================*
*=====================================================================*
*                                                                     *
*                      FSE-INTERCOM (Version 1.0)                     *
*                                                                     *
*                             May 1993                                *
*                                                                     *
*                           Martin Kropff                             *
*                                                                     *
* Department of Theoretical Production Ecology, P.O.B. 430, 6700 AK   *
* Wageningen, The Netherlands. Version September 1990                 *
*                                                                     *
*                                                                     *
* Simulation model for Crop-Weed Competition in FORTRAN, based on     *
* earlier versions, written in CSMP. References:                      *
* Kropff, M.J. (1988) Weed Research 28, 465-471                       *
* Kropff and Spitters (1992) Weed Research 32, 437-450.               *
*                                                                     *
* The model is programmed, using the FORTRAN Simulation Environment   *
* for Crop Growth Models (FSE), developed by D.W.G. van Kraalingen    *
* Simulation Report CABO-TT, no 23, July 1991, 77 pp.                 *
* Department of Theoretical Production Ecology and                    *
* Centre for Agrobiological Research, Wageningen, The Netherlands      *
*                                                                     *
* External files needed:   TIMER.DAT                                  *
*                          species.DAT (e.g. SUGARB.DAT; CHENO.DAT)   *
*                          SOIL.DAT                                    *
*                          weather files                              *
*                          RERUNS.DAT (only when reruns are needed)   *
*                                                                     *
*=====================================================================*
*=====================================================================*


        PROGRAM MAIN

*---- Standard declarations
      IMPLICIT REAL (A-Z)
      INTEGER      ITASK , INSETS, IRUN,   I1,     I2
      INTEGER      IUNITR, IUNITT, IUNITO, IUNITP, IUNITS
      INTEGER      ISTAT1, ISTAT2, IDAY,   IYEAR,  ISTN
      INTEGER      ITABLE, IDTMP,  IMNHD,  INHD
      LOGICAL      OUTPUT, TERMNL, WTRMES, WTROK
      CHARACTER*80 WTRDIR, FILER,  FILET,  FILEO,  FILES
      CHARACTER*7  CNTR
      CHARACTER*1  DUMMY
      PARAMETER (IMNHD=20)
```

```
      REAL HARDAY(IMNHD)
      PARAMETER (TINY=1.E-4)

*---- Special declaration required to write warnings to
*     output file
      COMMON /LOGCOM/ IUNITO, TIME, IRUN

*---- Unit numbers for rerun (R), timer (T), output (O),
*     plant data (P) and soil data (S) files.
*     WTRMES flags any messages from the weather system
      IUNITR = 20
      IUNITT = 30
      IUNITO = 40
      IUNITP = 50
      IUNITS = 60
      WTRMES = .FALSE.

*---- File names
      FILER  = 'RERUNS.DAT'
      FILET  = 'TIMER.DAT'
      FILEO  = 'RESULTS.OUT'
      FILES  = 'SOIL.DAT'

*---- Open output file, read number of rerun sets
      CALL FOPEN (IUNITO, FILEO, 'NEW', 'DEL')
      CALL COPFIL (IUNITT, FILET, IUNITO)

      CALL RDSETS (IUNITR, IUNITO, FILER, INSETS)
      IF (INSETS.GT.0) CALL COPFIL (IUNITR+1, FILER, IUNITO)


*========================================================================*
*                                                                        *
*                  Main loop and reruns begins here                      *
*                                                                        *
*========================================================================*

      DO 10 I1=0,INSETS

      IRUN = I1+1
      WRITE (*,'(A)') ' '

*---- Select data set
      CALL RDFROM (I1, .TRUE.)

*========================================================================*
*                                                                        *
*                     Initialization section                             *
*                                                                        *
*========================================================================*

      ITASK  = 1
      TERMNL = .FALSE.

*---- Read variables from TIMER.DAT file
      CALL RDINIT (IUNITT , IUNITO, FILET)
      CALL RDSCHA ('WTRDIR', WTRDIR)
      CALL RDSCHA ('CNTR'  , CNTR)
      CALL RDSREA ('DAYB'  , DAYB)
```

```
      CALL RDSREA ('FINTIM', FINTIM)
      CALL RDSREA ('PRDEL' , PRDEL)
      CALL RDSREA ('DELT'  , DELT)
      CALL RDSINT ('IYEAR' , IYEAR)
      CALL RDSINT ('ISTN'  , ISTN)
      CALL RDSINT ('ITABLE', ITABLE)
      CALL RDSINT ('IDTMP' , IDTMP)
      CALL RDAREA ('HARDAY', HARDAY, IMNHD, INHD)
      CLOSE (IUNITT, STATUS='DELETE')

*----- Initialize TIMER and OUTDAT routines
      CALL TIMER (ITASK, DAYB, DELT, PRDEL, FINTIM,
     &            IYEAR,  TIME, DAY, IDAY, TERMNL, OUTPUT)
      CALL OUTDAT (ITASK, IUNITO, 'TIME', TIME)

*----- Open weather file and read station information and return
*      weather data for start day of simulation
      CALL STINFO (1101, WTRDIR, ' ', CNTR, ISTN, IYEAR,
     &             ISTAT1, LONG, LAT, ELV, A, B)
      CALL WEATHR (IDAY, ISTAT2, DRAD, TMN, TMX, VAPOUR, WIND, RAIN)

*----- Conversions: Vapour pressure from kPa to mbar
*                   Total daily radiation from kJ/m2/d to J/m2/d
      VAPOUR = VAPOUR*10.
      AVRAD  = DRAD*1000.

      WTRMES = WTRMES .OR. (ISTAT1.NE.0) .OR. (ISTAT2.NE.0)
      WTROK  = (ISTAT1.EQ.0).AND.((ISTAT2.GE.0).OR.(ISTAT2.LT.-111111))
      TERMNL = TERMNL.OR..NOT.WTROK

*------< insert water balance call here if required >
      CALL WBAL (ITASK, IUNITS, IUNITO, FILES, OUTPUT, TERMNL,
     &           DAY,    DELT,
     &           RAIN, AEVAP, ATRAN,    RTD,
     &           VSMFC, VSMAD, VSMWP, VSMT, VSMRTZ)

*------< insert plant call here >
      CALL PLANTC (ITASK, IUNITP, IUNITO, FILET,OUTPUT,TERMNL,
     &             TIME,    DAY,   IDAY, DELT, LAT,
     &             AVRAD,   TMN,    TMX, VAPOUR, WIND,
     &             VSMWP, VSMFC, VSMRTZ,
     &             RTD,    FRD,    ES0, ATRAN)


*=======================================================================*
*                                                                       *
*                  Dynamic simulation section                          *
*                                                                       *
*=======================================================================*


20    IF (.NOT.TERMNL) THEN

      WRITE (*,'(A,I3,A,I5,A,F7.2)')
     & ' Run:', IRUN, ', Year:', IYEAR, ', Day:', DAY


*-----------------------------------------------------------------------*
*                  Integration of rates section                        *
*-----------------------------------------------------------------------*
```

```
      IF (ITASK.NE.1) THEN
      ITASK = 3

*-----< insert plant call here >
      CALL PLANTC (ITASK, IUNITP, IUNITO, FILET,OUTPUT,TERMNL,
     &                TIME,     DAY,    IDAY, DELT, LAT,
     &                AVRAD,    TMN,     TMX, VAPOUR, WIND,
     &                VSMWP,  VSMFC, VSMRTZ,
     &                 RTD,    FRD,     ES0, ATRAN)

*-----< insert water balance call here if required >
      CALL WBAL (ITASK, IUNITS, IUNITO, FILES, OUTPUT, TERMNL,
     &                DAY,    DELT,
     &                RAIN,  AEVAP,  ATRAN,    RTD,
     &                VSMFC, VSMAD,  VSMWP, VSMT, VSMRTZ)
      END IF

      ITASK = 2


*----------------------------------------------------------------------*
*              Calculation of driving variables section                *
*----------------------------------------------------------------------*

*---- Open weather file
      CALL STINFO (1101, WTRDIR, ' ', CNTR, ISTN, IYEAR,
     &                ISTAT1, LONG, LAT, ELV, A, B)
      CALL WEATHR (IDAY, ISTAT2, DRAD, TMN, TMX, VAPOUR, WIND, RAIN)

      IF (OUTPUT.OR.TERMNL) THEN
         CALL OUTDAT (ITASK, IUNITO, 'TIME', TIME)
         CALL OUTDAT (ITASK, IUNITO, 'DAY' , DAY)
      END IF

*---- Conversions: Vapour pressure from kPa to mbar
*                  Total daily radiation from kJ/m2/d to J/m2/d
      VAPOUR = VAPOUR*10.
      AVRAD  = DRAD*1000.

      WTRMES = WTRMES .OR. (ISTAT1.NE.0) .OR. (ISTAT2.NE.0)
      WTROK  = (ISTAT1.EQ.0).AND.((ISTAT2.GE.0).OR.
     &         (ISTAT2.LT.-111111))
      TERMNL = TERMNL.OR..NOT.WTROK


*----------------------------------------------------------------------*
*                  Calculation of rates section                        *
*----------------------------------------------------------------------*

*-----< insert plant call here >
      CALL PLANTC (ITASK, IUNITP, IUNITO, FILET,OUTPUT,TERMNL,
     &                TIME,     DAY,    IDAY, DELT, LAT,
     &                AVRAD,    TMN,     TMX, VAPOUR, WIND,
     &                VSMWP,  VSMFC, VSMRTZ,
     &                 RTD,    FRD,     ES0, ATRAN)

*-----< insert potential soil evaporation call here if required >
      CALL DEVAP (VSMT, VSMFC, VSMAD, ES0, FRD,
     &                AEVAP)
```

```
*-----< insert water balance call here if required >
      CALL WBAL (ITASK, IUNITS, IUNITO, FILES, OUTPUT, TERMNL,
     &                DAY,   DELT,
     &                RAIN,  AEVAP,  ATRAN,   RTD,
     &                VSMFC, VSMAD,  VSMWP,  VSMT, VSMRTZ)

*---- Time update, check for FINTIM and OUTPUT
      CALL TIMER (ITASK, DAYB, DELT, PRDEL, FINTIM,
     &                IYEAR,  TIME, DAY, IDAY, TERMNL, OUTPUT)

*---- Generate output to file if day is equal to a harvest day
      IF (HARDAY(1).NE.0.) THEN
          DO 30 I2=1,INHD
             IF (DAY.GT.(HARDAY(I2)-TINY).AND.DAY.LT.(HARDAY(I2)+TINY))
     &          OUTPUT = .TRUE.
30        CONTINUE
      ELSE IF (INHD.GT.1) THEN
          CALL ERROR ('FSE-MAIN','harvest data in TIMER.DAT not correct')
      END IF

      GOTO 20
      END IF


*========================================================================*
*                                                                        *
*                         Terminal section                               *
*                                                                        *
*========================================================================*

      ITASK = 4

*---- Generate output file dependent on option from timer file
      IF (ITABLE.GE.4) CALL OUTDAT (ITABLE, 20, ' ',0.)

*-----< insert plant call here >
      CALL PLANTC (ITASK, IUNITP, IUNITO, FILET,OUTPUT,TERMNL,
     &                TIME,    DAY,   IDAY, DELT, LAT,
     &                AVRAD,    TMN,    TMX, VAPOUR, WIND,
     &                VSMWP,  VSMFC, VSMRTZ,
     &                  RTD,    FRD,    ES0, ATRAN)

*-----< insert water balance call here if required >
      CALL WBAL (ITASK, IUNITS, IUNITO, FILES, OUTPUT, TERMNL,
     &                DAY,   DELT,
     &                RAIN,  AEVAP,  ATRAN,   RTD,
     &                VSMFC, VSMAD,  VSMWP,  VSMT, VSMRTZ)

*---- Delete temporary output file dependent on switch from timer file
      IF (IDTMP.EQ.1) CALL OUTDAT (99, 0, ' ', 0.)

10    CONTINUE

*---- Delete temporary rerun file if reruns were carried out
      IF (INSETS.GT.0) CLOSE (IUNITR, STATUS='DELETE')

      IF (WTRMES) THEN
         WRITE (*,'(A,/,A)')
     &     ' There have been errors and/or warnings from',
```

```
      &      ' the weather system, check file WEATHER.LOG'
             WRITE (IUNITO,'(A,/,A)')
      &        ' There have been errors and/or warnings from',
      &        ' the weather system, check file WEATHER.LOG'
             WRITE (*,'(A)') ' Press <RETURN>'
             READ (*,'(A)') DUMMY
           END IF

           STOP
           END
```

```
*=======================================================================*
*=======================================================================*
*                                                                       *
* SUBROUTINE PLANTC                                                     *
* Date    : May 1993                                                   *
*                                                                       *
* Version: 1.0                                                          *
* Purpose: This subroutine simulates growth of competing species in     *
*          potential and water limited production situations.           *
*                                                                       *
*                                                                       *
* FORMAL PARAMETERS: (I=input, O=output, C=control, IN=init, T=time)    *
* name    type description                              units   class  *
* ----    ---- -----------                              -----   -----  *
* ITASK   I4   determines action of the subroutine,       -      C,I   *
*              1=initialization, 2=rate calculation,                    *
*              3=integration, 4=terminal                                *
* IUNITP  I4   unit number of plant data file             -      C,I   *
* IUNITO  I4   unit number of output file                 -      C,I   *
* FILET   C*   file name for time variables               -      C,I   *
* OUTPUT  L4   flag that indicates if output to file is   -      C,I   *
*              required                                                  *
* TERMNL  L4   flag that indicates if simulation should   -    C,I,O   *
*              terminate                                                 *
* TIME    R4   daynumber start simulation                 d      T,I   *
* DATE    R4   daynumber since 1 January                  d      T     *
* IDAY    I4   integer variable for DATE                  d      T     *
* DELT    R4   time interval of integration               d      T     *
* LAT     R4   latitude of weather station              degrees  I     *
* AVRAD   R4   daily incoming total global radiation     J/m2/d   I     *
* TMN     R4   daily minimum temperature         degrees Celsius  I     *
* TMX     R4   daily maximum temperature         degrees Celsius  I     *
* VAPOUR  R4   average vapour pressure                   mbar     I     *
* WIND    R4   daily average wind speed                  m/s      I     *
* VSMWP   R4   soil moisture content at wilting point    m3/m3    I     *
* VSMFC   R4   soil moisture content at field capacity   m3/m3    I     *
* VSMRTZ  R4   actual soil moisture content root layer   m3/m3    I     *
* RTD     R4   rooted depth                              m        I     *
* FRD     R4   fraction global radiation used for drying  -       I     *
*              power in Penman evaporation                               *
* ES0     R4   Penman evaporation of a bare soil         mm/d     I     *
* ATRAN   R4   total actual transpiration rate canopy    mm/d     I     *
*                                                                       *
* FATAL ERROR CHECKS (execution terminated, message)                   *
* DELT < 1.0                                                            *
* Certain sequences of ITASK, see subroutine CHKTSK                     *
* INS .NE. INS1 or INS .NE. INS2                                        *
* FLV(IS)+FST(IS)+FSTOA(IS) > 1.                                        *
```

```
*                                                                      *
* SUBROUTINES and FUNCTIONS called: CHKTSK, OUTCOM, ERROR, RDINIT,     *
*      RDAREA, RDSREA, COPFIL, ASTRO, TOTASS, TOTRAN, OUTARR, OUTDAT   *
*      OUTPLT, LINT, INTGRL                                            *
*                                                                      *
* FILE usage: - time variables file FILET                             *
*             - plant data file with unit IUNITP                      *
*             - output file with unit IUNITO for output and warnings  *
*                                                                      *
*======================================================================*
*======================================================================*


        SUBROUTINE PLANTC (ITASK, IUNITP,IUNITO, FILET,
     &                     OUTPUT,TERMNL,
     &                     TIME,  DATE,   IDAY,   DELT,   LAT,
     &                     AVRAD, TMN,    TMX,    VAPOUR, WIND,
     &                     VSMWP, VSMFC,  VSMRTZ,
     &                     RTD,   FRD,    ES0,    ATRAN)

        IMPLICIT REAL (A-Z)

*---- Formal parameters
        INTEGER   ITASK, IUNITP, IUNITO, IDAY
        LOGICAL   OUTPUT, TERMNL
        CHARACTER*(*) FILET

*---- Standard local declarations
        INTEGER ITOLD, IMNS,IMNP, INS,INS1,INS2, IS, IRUNLA
        PARAMETER (IMNS=10,IMNP=40)
        INTEGER IPSPEC(IMNS), IDAYEM(IMNS)
        REAL NPL(IMNS)
        LOGICAL IN1TC,INITS

*---- States
        REAL LA0(IMNS),   WLVGI(IMNS), WSTGI(IMNS), WRTI(IMNS)
        REAL WLVG(IMNS),  WLVGM(IMNS), WSTG(IMNS),  WRT(IMNS)
        REAL HGHTI(IMNS), HGHT(IMNS),  WLVD(IMNS),  WSOD(IMNS)
        REAL WSTD(IMNS),  WSOA(IMNS),  WSOB(IMNS)
        REAL TS(IMNS),    TSLV(IMNS)
        REAL WAG(IMNS),   WTOT(IMNS),  WDTOT(IMNS)
        REAL LAI(IMNS),   LAIM(IMNS),  SAI(IMNS),  FAI(IMNS)

*---- Species parameters and rates
        INTEGER ILREDF(IMNS), ILREDM(IMNS)
        INTEGER ILAMD(IMNS),  ILEFF(IMNS)
        REAL AMAX(IMNS),  AMAXM(IMNS) , AMAXS(IMNS), AMAXF(IMNS)
        REAL AMDVS(IMNS), AMDVST(IMNP,IMNS)
        REAL REDF(IMNS),  REDFTB(IMNP,IMNS)
        REAL REDMN(IMNS), REDMNT(IMNP,IMNS)
        REAL EFF(IMNS),   EFFTB(IMNP,IMNS)

        REAL KDF(IMNS), KS(IMNS), KF(IMNS)

        REAL TMD(IMNS), TMDLV(IMNS), TMXD(IMNS), TMXLV(IMNS)

        INTEGER ILPTB(IMNS)
        REAL PTB(IMNP,IMNS)
        REAL CRPF(IMNS)
```

```
      REAL HU(IMNS),         HULV(IMNS)

      REAL FRABS(IMNS),  GPHOT(IMNS),  DTGA(IMNS)
      REAL MAINTS(IMNS), MAINLV(IMNS), MAINST(IMNS), MNDVS(IMNS)
      REAL MAINSB(IMNS), MAINRT(IMNS), MAINT(IMNS),  MAINSA(IMNS)

      INTEGER ILFAG(IMNS),ILFLV(IMNS), ILFST(IMNS),   ILFRT(IMNS)
      INTEGER ILFSOA(IMNS)
      REAL FAG(IMNS),        FAGTB(IMNP,IMNS)
      REAL FLV(IMNS),        FLVTB(IMNP,IMNS)
      REAL FST(IMNS),        FSTTB(IMNP,IMNS)
      REAL FRT(IMNS),        FRTTB(IMNP,IMNS)
      REAL FSOA(IMNS),       FSOATB(IMNP,IMNS)
      REAL FSOB(IMNS)

      REAL ASRQ(IMNS),       ASRQLV(IMNS),      ASRQST(IMNS)
      REAL ASRQRT(IMNS),     ASRQSB(IMNS),      ASRQSA(IMNS)
      REAL GTW(IMNS)

      INTEGER ILRDRS(IMNS), ILRDRO(IMNS)
      REAL DRL(IMNS),        TSLAM(IMNS)
      REAL RDRST(IMNS),      RDRSTT(IMNP,IMNS)
      REAL RDRSOA(IMNS),     RDRSOT(IMNP,IMNS)

      REAL REDIST(IMNS),  REDLM(IMNS)

      REAL YLV(IMNS),   YST(IMNS), DLV(IMNS), DST(IMNS), DSOA(IMNS)
      REAL REDST(IMNS),GAG(IMNS), GLV(IMNS), GRLV(IMNS),GRST(IMNS)
      REAL GRSOA(IMNS),GRT(IMNS), GSOB(IMNS)

      INTEGER ILLAI(IMNS)
      REAL  LAITB(IMNP, IMNS)

      INTEGER ILSLA(IMNS)
      REAL SLA(IMNS), SLATB(IMNP,IMNS)
      REAL SSA(IMNS), SFA(IMNS), RGRL(IMNS)
      REAL GLAI(IMNS), LAIY(IMNS),  LAID(IMNS)
      REAL SSL(IMNS) , SSLMAX(IMNS),AS(IMNS),   BS(IMNS)
      REAL RHG(IMNS), HMAX(IMNS),  HS(IMNS),   HB(IMNS)

      REAL TRAN(IMNS),  TRANRF(IMNS)

      SAVE

      DATA ITOLD /4/,INITS /.FALSE./,INITC/.FALSE./

*     The task that the subroutine should do (ITASK) against the
*     task that was done during the previous call (ITOLD) is
*     checked. Only certain combinations are allowed. These are:
*
*         New task:             Old task:
*         initialization        terminal
*         integration           rate calculation
*         rate calculation      initialization, integration
*         terminal              <any old task>
*     Note: there is one combination that is correct but will not
*     cause calculations to be done i.e. if integration is required
*     immediately after initialization.
```

```
      CALL CHKTSK ('PLANTC', IUNITO, ITOLD, ITASK)

*=======================================================================*
*                         Initialization                                *
*=======================================================================*

      IF (ITASK.EQ.1) THEN

*------- Send title to output file
      CALL OUTCOM ('FSE-INTERCOM: Competition model')

*------- Initialization section

      IF (DELT.LT.1.0) CALL ERROR
     &     ('PLANTC','DELT too small for PLANTC')

      CALL RDINIT (IUNITP, IUNITO, FILET)

*------- Initialization of states

      CALL RDAINT ('IPSPEC', IPSPEC, IMNS, INS)

      CALL RDAINT ('IDAYEM', IDAYEM, IMNS, INS1)

      CALL RDAREA ('NPL', NPL, IMNS, INS2)
      CALL RDSINT ('IRUNLA',IRUNLA)

      IF (IRUNLA.EQ.1) CALL OUTCOM ('Measured LAI as input !')
      IF (IRUNLA.EQ.0) CALL OUTCOM ('LAI simulated !')

      CLOSE (IUNITP, STATUS='DELETE')
      IF (INS.NE.INS1.OR.INS.NE.INS2) CALL ERROR('PLANTC',
     &     'Inconsistent initialization in TIMER.DAT')

      LAITOT = 0.
      RTD    = 0.

*------------- Initialization of species parameters

      DO 30 IS=1,INS

         IF (IPSPEC(IS).EQ.1) THEN

            IF (.NOT.INITS) THEN
               CALL COPFIL (IUNITP, 'SUGARB.DAT', IUNITO)
               INITS = .TRUE.
            ENDIF
            CALL RDINIT (IUNITP, IUNITO, 'SUGARB.DAT')

         ELSE IF (IPSPEC(IS).EQ.2) THEN
            IF (.NOT.INITC) THEN
               CALL COPFIL (IUNITP, 'CHENO.DAT', IUNITO)
               INITC = .TRUE.
            ENDIF

            CALL RDINIT (IUNITP, IUNITO, 'CHENO.DAT')

         ELSE
```

```
              CALL ERROR ('PLANTC','unknown species')

          END IF

*---------- Sugar beet initialization

          IF (IPSPEC(IS).EQ.1) THEN

          CALL OUTCOM ('Sugar beet (1)')

*------------- States
              CALL RDSREA ('LA0', LA0(IS))
              CALL RDSREA ('WLVGI', WLVGI(IS))
              CALL RDSREA ('WSTGI', WSTGI(IS))
              CALL RDSREA ('WRTI' , WRTI(IS))
              CALL RDSREA ('HGHTI' , HGHTI(IS))
              LAI(IS)   = 0.
              SAI(IS)   = 0.
              FAI(IS)   = 0.
              WLVG(IS)  = 0.
              WLVD(IS)  = 0.
              WSTG(IS)  = 0.
              WSTD(IS)  = 0.
              WRT(IS)   = 0.
              HGHT(IS)  = 0.
              WSOA(IS)  = 0.
              WSOD(IS)  = 0.
              WSOB(IS)  = 0.
              TS(IS)    = 0.
              TSLV(IS)  = 0.
              SSL(IS)   = 0.
              SSLMAX(IS) = 0.
              WAG(IS)   = WLVG(IS)+WSTG(IS)+WSOA(IS)
              WTOT(IS)  = WAG(IS)  +WSOB(IS)
              WDTOT(IS)= WTOT(IS)+WLVD(IS)+WSTD(IS)+WSOD(IS)

*------------- Rates
              GLAI(IS)  = 0.
              DTGA(IS)  = 0.
              MAINTS(IS)= 0.
              RHG(IS)   = 0.
              GTW(IS)   = 0.

*------------- Other parameters

              CALL RDSREA ('RTD',RTDT)
              IF (RTDT.GT.RTD) RTD = RTDT

              CALL RDSREA ('AMAXM',AMAXM(IS))
              CALL RDSREA ('AMAXS',AMAXS(IS))
              CALL RDSREA ('AMAXF',AMAXF(IS))
              CALL RDAREA ('EFFTB',EFFTB(1,IS),IMNP,ILEFF(IS))
              CALL RDSREA ('KDF', KDF(IS))
              CALL RDSREA ('KS',  KS(IS))
              CALL RDSREA ('KF',  KF(IS))
              CALL RDSREA ('CRPF',CRPF(IS))
              CALL RDAREA ('PTB', PTB(1,IS),IMNP,ILPTB(IS))
              CALL RDAREA ('REDFTB',REDFTB(1,IS),IMNP,ILREDF(IS))
```

```
          CALL RDAREA ('REDMNT',REDMNT(1,IS),IMNP,ILREDM(IS))
          CALL RDAREA ('AMDVST',AMDVST(1,IS),IMNP,ILAMD(IS))
          CALL RDSREA ('TMD',    TMD(IS))
          CALL RDSREA ('TMDLV', TMDLV(IS))
          CALL RDSREA ('TMXD',  TMXD(IS))
          CALL RDSREA ('TMXLV', TMXLV(IS))
          CALL RDSREA ('MAINLV',MAINLV(IS))
          CALL RDSREA ('MAINST',MAINST(IS))
          CALL RDSREA ('MAINSA',MAINSA(IS))
          CALL RDSREA ('MAINSB',MAINSB(IS))
          CALL RDSREA ('MAINRT',MAINRT(IS))
          CALL RDAREA ('FAGTB', FAGTB(1,IS),  IMNP,ILFAG(IS))
          CALL RDAREA ('FLVTB', FLVTB(1,IS),  IMNP,ILFLV(IS))
          CALL RDAREA ('FSTTB', FSTTB(1,IS),  IMNP,ILFST(IS))
          CALL RDAREA ('FRTTB', FRTTB(1,IS),  IMNP,ILFRT(IS))
          CALL RDAREA ('FSOATB',FSOATB(1,IS),IMNP,ILFSOA(IS))
          CALL RDSREA ('ASRQLV', ASRQLV(IS))
          CALL RDSREA ('ASRQST', ASRQST(IS))
          CALL RDSREA ('ASRQSA', ASRQSA(IS))
          CALL RDSREA ('ASRQRT', ASRQRT(IS))
          CALL RDSREA ('ASRQSB', ASRQSB(IS))
          CALL RDSREA ('DRL' ,    DRL(IS))
          CALL RDSREA ('TSLAM' , TSLAM(IS))

          CALL RDAREA ('RDRSTT',RDRSTT(1,IS),IMNP,ILRDRS(IS))
          CALL RDSREA ('REDLM', REDLM(IS))
          CALL RDSREA ('REDST', REDST(IS))

          CALL RDAREA ('RDRSOT',RDRSOT(1,IS),IMNP,ILRDRO(IS))
          CALL RDAREA ('LAITB', LAITB(1,IS),  IMNP,ILLAI(IS))
          CALL RDAREA ('SLATB', SLATB(1,IS),  IMNP,ILSLA(IS))
          CALL RDSREA ('SSA', SSA(IS))
          CALL RDSREA ('SFA', SFA(IS))
          CALL RDSREA ('RGRL',RGRL(IS))
          CALL RDSREA ('HMAX',HMAX(IS))
          CALL RDSREA ('HS',  HS(IS))
          CALL RDSREA ('HB',  HB(IS))

          CLOSE (IUNITP, STATUS='DELETE')

        END IF

*---------- Chenopodium album initialization

        IF (IPSPEC(IS).EQ.2) THEN

        CALL OUTCOM ('Chenopodium album (2)')

*------------- States
          CALL RDSREA ('LA0',    LA0(IS))
          CALL RDSREA ('WLVGI', WLVGI(IS))
          CALL RDSREA ('WSTGI', WSTGI(IS))
          CALL RDSREA ('WRTI' , WRTI(IS))
          CALL RDSREA ('HGHTI', HGHTI(IS))
          LAI(IS)  = 0.
          SAI(IS)  = 0.
          FAI(IS)  = 0.
          WLVG(IS) = 0.
          WLVD(IS) = 0.
```

```
                WSTG(IS)  = 0.
                WSTD(IS)  = 0.
                WRT(IS)   = 0.
                HGHT(IS)  = 0.
                WSOA(IS)  = 0.
                WSOD(IS)  = 0.
                WSOB(IS)  = 0.
                TS(IS)    = 0.
                TSLV(IS)  = 0.
                SSL(IS)   = 0.
                SSLMAX(IS) = 0.
                WAG(IS)   = WLVG(IS)+WSTG(IS)+WSOA(IS)
                WTOT(IS)  = WAG(IS)  +WSOB(IS)
                WDTOT(IS) = WTOT(IS)+WLVD(IS)+WSTD(IS)+WSOD(IS)

*------------- Rates
                GLAI(IS)  = 0.
                DTGA(IS)  = 0.
                MAINTS(IS)= 0.
                MAINT(IS) = 0.
                RHG(IS)   = 0.
                GTW(IS)   = 0.

*------------- Other parameters

                CALL RDSREA ('RTD',RTDT)
                IF (RTDT.GT.RTD) RTD = RTDT

                CALL RDSREA ('AMAXM',AMAXM(IS))
                CALL RDSREA ('AMAXS',AMAXS(IS))
                CALL RDSREA ('AMAXF',AMAXF(IS))
                CALL RDAREA ('EFFTB',EFFTB(1,IS),IMNP,ILEFF(IS))
                CALL RDSREA ('KDF',   KDF(IS))
                CALL RDSREA ('KS',    KS(IS))
                CALL RDSREA ('KF',    KF(IS))
                CALL RDSREA ('CRPF', CRPF(IS))
                CALL RDAREA ('PTB',    PTB(1,IS),    IMNP,ILPTB(IS))
                CALL RDAREA ('REDFTB',REDFTB(1,IS),IMNP,ILREDF(IS))
                CALL RDAREA ('REDMNT',REDMNT(1,IS),IMNP,ILREDM(IS))
                CALL RDAREA ('AMDVST',AMDVST(1,IS),IMNP,ILAMD(IS))
                CALL RDSREA ('TMD',    TMD(IS))
                CALL RDSREA ('TMDLV', TMDLV(IS))
                CALL RDSREA ('TMXD',   TMXD(IS))
                CALL RDSREA ('TMXLV', TMXLV(IS))
                CALL RDSREA ('MAINLV',MAINLV(IS))
                CALL RDSREA ('MAINST',MAINST(IS))
                CALL RDSREA ('MAINSA',MAINSA(IS))
                CALL RDSREA ('MAINSB',MAINSB(IS))
                CALL RDSREA ('MAINRT',MAINRT(IS))
                CALL RDAREA ('FAGTB', FAGTB(1,IS), IMNP,ILFAG(IS))
                CALL RDAREA ('FLVTB', FLVTB(1,IS), IMNP,ILFLV(IS))
                CALL RDAREA ('FSTTB', FSTTB(1,IS), IMNP,ILFST(IS))
                CALL RDAREA ('FRTTB', FRTTB(1,IS), IMNP,ILFRT(IS))
                CALL RDAREA ('FSOATB',FSOATB(1,IS),IMNP,ILFSOA(IS))
                CALL RDSREA ('ASRQLV', ASRQLV(IS))
                CALL RDSREA ('ASRQST', ASRQST(IS))
                CALL RDSREA ('ASRQSA', ASRQSA(IS))
                CALL RDSREA ('ASRQRT', ASRQRT(IS))
                CALL RDSREA ('ASRQSB', ASRQSB(IS))
```

```
          CALL RDSREA ('DRL' ,    DRL(IS))
          CALL RDSREA ('TSLAM' , TSLAM(IS))


          CALL RDAREA ('RDRSTT',RDRSTT(1,IS),IMNP,ILRDRS(IS))
          CALL RDSREA ('REDLM', REDLM(IS))
          CALL RDSREA ('REDST', REDST(IS))


          CALL RDAREA ('RDRSOT',RDRSOT(1,IS),IMNP,ILRDRO(IS))
          CALL RDAREA ('LAITB', LAITB(1,IS), IMNP,ILLAI(IS))
          CALL RDAREA ('SLATB', SLATB(1,IS), IMNP,ILSLA(IS))
          CALL RDSREA ('SSA', SSA(IS))
          CALL RDSREA ('SFA', SFA(IS))
          CALL RDSREA ('RGRL',RGRL(IS))
          CALL RDSREA ('HMAX',HMAX(IS))
          CALL RDSREA ('HS',  HS(IS))
          CALL RDSREA ('HB',  HB(IS))
          CALL RDSREA ('AS',  AS(IS))
          CALL RDSREA ('BS',  BS(IS))


          CLOSE (IUNITP, STATUS='DELETE')

        END IF

*        IF (IPSPEC(IS).EQ.3) THEN
*          Initialization of new species
*          CALL RDINIT (IUNITP, IUNITO, 'SPEC.DAT')
*          CLOSE (IUNITP, STATUS='DELETE')
*        END IF

30     CONTINUE


*=======================================================================*
*                   Rate calculation section                           *
*=======================================================================*

*------- Rate calculations

      ELSE IF (ITASK.EQ.2) THEN

*------- Driving variables
        TMPA  = 0.5 * (TMX  + TMN)
        TMTMX = 0.5 * (TMPA + TMX)

        CALL ASTRO (DATE,LAT,
     &              DAYL,DAYLP,SINLD,COSLD)

*------- Calculation photosynthesis parameters and reduction factors

        DO 40 IS=1,INS

*---------- Sugar beet

          IF (IPSPEC(IS).EQ.1) THEN
            EFF(IS)   = LINT (EFFTB(1,IS),ILEFF(IS),TMTMX)
            REDF(IS)  = LINT (REDFTB(1,IS),ILREDF(IS),TMTMX)
            REDMN(IS) = LINT (REDMNT(1,IS),ILREDM(IS),TMN)
            AMDVS(IS) = LINT (AMDVST(1,IS),ILAMD(IS),TS(IS))
            AMAX(IS)  = AMAXM(IS)*REDF(IS)*AMDVS(IS)*REDMN(IS)
```

```
          ENDIF

*---------- Chenopodium album

          IF (IPSPEC(IS).EQ.2) THEN
              EFF(IS)    = LINT (EFFTB(1,IS),ILEFF(IS),TMTMX)
              REDF(IS)   = LINT (REDFTB(1,IS),ILREDF(IS),TMTMX)
              REDMN(IS)  = LINT (REDMNT(1,IS),ILREDM(IS),TMN)
              AMDVS(IS)  = LINT (AMDVST(1,IS),ILAMD(IS),TS(IS))
              AMAX(IS)   = AMAXM(IS)*REDF(IS)*AMDVS(IS)*REDMN(IS)
              AMAXS(IS)  = AMAXS(IS)*REDF(IS)*AMDVS(IS)*REDMN(IS)
              AMAXF(IS)  = AMAXF(IS)*REDF(IS)*AMDVS(IS)*REDMN(IS)
          ENDIF

*---------- Other species

*         IF (IPSPEC(IS).EQ.3) THEN
*
*         ENDIF

40        CONTINUE

*------- Calculation of light absorption and photosynthesis of
*        competing species

          CALL TOTASS (DATE,DAYL,KDF,KS,KF,AMAX,AMAXS,AMAXF,EFF,
     &                 LAI,SAI,FAI,AVRAD,SINLD,COSLD,INS,HGHT,
     &                 ATMTR,FRABS,FRD,DTGA)

          CALL TOTRAN (INS,TMN,TMX,AVRAD,ATMTR,WIND,VAPOUR,
     &                 CRPF,VSMFC,VSMWP,VSMRTZ,FRABS,PTB,ILPTB,
     &                 E0,ES0,ET0,TRAN,ATRAN,TRANRF)

*----- Calculation of growth rates, development rates

        DO 50 IS=1,INS

*-------- Sugar beet

        IF (IPSPEC(IS).EQ.1) THEN

           IF (IDAY.GE.IDAYEM(IS)) THEN

              HU(IS)    = MIN(TMXD(IS)-TMD(IS),(MAX (0., TMPA-TMD(IS))))
              HULV(IS)  = MIN(TMXLV(IS)-TMDLV(IS),(MAX (0., TMPA-TMDLV(IS))))
              GPHOT(IS) = DTGA(IS)*30./44.
              Q10       = 2.
              REFTMP    = 25.
              TEFF      = Q10**((TMPA-REFTMP)/10.)

               IF ((WLVG(IS)+WLVD(IS)).GT.0.) THEN
                 MNDVS(IS) = WLVG(IS)/(WLVG(IS)+WLVD(IS))
               ELSE
                 MNDVS(IS) = 1.
               ENDIF

              MAINTS(IS) = WLVG(IS)*MAINLV(IS)+WSTG(IS)*MAINST(IS)+
     &                     WSOA(IS)*MAINSA(IS)+WSOB(IS)*MAINSB(IS)+
     &                     WRT(IS) *MAINRT(IS)
```

```
        MAINT(IS)   = MIN (GPHOT(IS),MAINTS(IS)*TEFF*
    &                        MNDVS(IS))
        FAG(IS)     = LINT (FAGTB(1,IS),   ILFAG(IS),  TS(IS))
        FLV(IS)     = LINT (FLVTB(1,IS),   ILFLV(IS),  TS(IS))
        FST(IS)     = LINT (FSTTB(1,IS),   ILFST(IS),  TS(IS))
        FSOA(IS)    = LINT (FSOATB(1,IS),  ILFSOA(IS), TS(IS))
        FRT(IS)     = LINT (FRTTB(1,IS),   ILFRT(IS),  TS(IS))
        FSOB(IS)    = 1.-(FLV(IS)+FST(IS)+FSOA(IS))

        IF ((FLV(IS)+FST(IS)+FSOA(IS)).GT.1.)
    &   CALL ERROR ('PLANTC','fractions partitioning >1')

        ASRQ(IS)    = FAG(IS)*(ASRQLV(IS)*FLV(IS)+
    &                   ASRQST(IS)*FST(IS)+
    &                   ASRQSA(IS)*FSOA(IS)+ASRQSB(IS)*FSOB(IS))+
    &                   ASRQRT(IS)*FRT(IS)

        RDRST(IS) = LINT (RDRSTT(1,IS), ILRDRS(IS), TS(IS))

        IF (TS(IS).GT.TSLAM(IS).AND.WLVG(IS).GT.0.) THEN
          YLV(IS) = -WLVGM(IS)*DRL(IS)*HU(IS)
        ELSE
          YLV(IS) = 0.
        ENDIF

        YST(IS)     = WSTG(IS)*(EXP (RDRST(IS)*HU(IS))-1.)
        DLV(IS)     = (1.-REDLM(IS))*YLV(IS)
        DST(IS)     = (1.-REDST(IS))*YST(IS)
        REDIST(IS)  = REDLM(IS)*YLV(IS) + REDST(IS)*YST(IS)
        RDRSOA(IS)  = LINT (RDRSOT(1,IS),ILRDRO(IS),TS(IS))
        DSOA(IS)    = WSOA(IS)*RDRSOA(IS)

        GTW(IS)     = (GPHOT(IS)-MAINT(IS)+REDIST(IS))*
    &                   TRANRF(IS)/ASRQ(IS)

        GAG(IS)     = GTW(IS)*FAG(IS)
        GRT(IS)     = GTW(IS)*FRT(IS)
        GLV(IS)     = GAG(IS)*FLV(IS)
        GRLV(IS)    = GLV(IS)-YLV(IS)
        GRST(IS)    = GAG(IS)*FST(IS)-YST(IS)
        GRSOA(IS)   = GAG(IS)*FSOA(IS)-DSOA(IS)
        GSOB(IS)    = GAG(IS)*FSOB(IS)

    IF (IRUNLA.EQ.1) THEN
        LAID(IS) = LINT (LAITB(1,IS), ILLAI(IS), DATE)
        LAIY(IS) = LINT (LAITB(1,IS), ILLAI(IS), (DATE-1.))
        GLAI(IS) = LAID(IS)-LAIY(IS)
    ELSE
        SLA(IS) = LINT (SLATB(1,IS), ILSLA(IS), TS(IS))
        GLAI(IS) = SLA(IS)*GRLV(IS)
          IF (LAITOT.LT.0.75) THEN
          GLAI(IS) = LAI(IS)*(EXP (RGRL(IS)*HULV(IS))-1.)
          IF (IDAY.EQ.IDAYEM(IS))
    &     GLAI(IS) = LA0(IS)*NPL(IS)*1.E-4
          ENDIF
        IF (TS(IS).GT.TSLAM(IS).AND.LAI(IS).GT.0.)
    &                   GLAI(IS)= LAIM(IS)*DRL(IS)*HU(IS)
    ENDIF
```

```
           RHG(IS) = HU(IS)*HMAX(IS)*HS(IS)*HB(IS)*
     &                EXP(-HS(IS)*TS(IS))/
     &                (1.+HB(IS)*EXP(-HS(IS)*TS(IS)))**2*
     &                TRANRF(IS)

        ENDIF

     ENDIF

*-------- Chenopodium album

     IF (IPSPEC(IS).EQ.2) THEN

      IF (IDAY.GE.IDAYEM(IS)) THEN

        HU(IS)     = MIN(TMXD(IS)-TMD(IS),(MAX (0., TMPA-TMD(IS))))
        HULV(IS)   = MIN(TMXLV(IS)-TMDLV(IS),(MAX (0., TMPA-TMDLV(IS))))
        GPHOT(IS)  = DTGA(IS)*30./44.
        Q10        = 2.
        REFTMP     = 25.
        TEFF       = Q10**((TMPA-REFTMP)/10.)

         IF ((WLVG(IS)+WLVD(IS)).GT.0.) THEN
             MNDVS(IS) = WLVG(IS)/(WLVG(IS)+WLVD(IS))
         ELSE
             MNDVS(IS) = 1.
         ENDIF

        MAINTS(IS) = WLVG(IS)*MAINLV(IS)+WSTG(IS)*MAINST(IS)+
     &                WSOA(IS)*MAINSA(IS)+WSOB(IS)*MAINSB(IS)+
     &                WRT(IS)*MAINRT(IS)
        MAINT(IS)  = MIN (GPHOT(IS), MAINTS(IS)*TEFF*
     &                                 MNDVS(IS))

        FAG(IS)    = LINT (FAGTB(1,IS), ILFAG(IS), TS(IS))
        FLV(IS)    = LINT (FLVTB(1,IS), ILFLV(IS), TS(IS))
        FST(IS)    = LINT (FSTTB(1,IS), ILFST(IS), TS(IS))
        FSOA(IS)   = LINT (FSOATB(1,IS), ILFSOA(IS), TS(IS))
        FRT(IS)    = LINT (FRTTB(1,IS), ILFRT(IS), TS(IS))
        FSOB(IS)   = 1.-(FLV(IS)+FST(IS)+FSOA(IS))
        ASRQ(IS)   = FAG(IS)*(ASRQLV(IS)*FLV(IS)
     &                +ASRQST(IS)*FST(IS)
     &                +ASRQSA(IS)*FSOA(IS)+ASRQSB(IS)*FSOB(IS))
     &                +ASRQRT(IS)*FRT(IS)

        RDRST(IS)  = LINT (RDRSTT(1,IS), ILRDRS(IS), TS(IS))

        IF (TS(IS).GT.TSLAM(IS).AND.WLVG(IS).GT.0.) THEN
           YLV(IS)  = -WLVGM(IS)*DRL(IS)*HU(IS)
        ELSE
           YLV(IS)  = 0.
        ENDIF

        YST(IS)    = WSTG(IS)*(EXP (RDRST(IS)*HU(IS))-1.)
        DLV(IS)    = (1.-REDLM(IS))*YLV(IS)
        DST(IS)    = (1.-REDST(IS))*YST(IS)
        REDIST(IS) = REDLM(IS)*YLV(IS) + REDST(IS)*YST(IS)
        RDRSOA(IS) = LINT (RDRSOT(1,IS),ILRDRO(IS),TS(IS))
        DSOA(IS)   = WSOA(IS)*RDRSOA(IS)
```

```
         GTW(IS)     = (GPHOT(IS)-MAINT(IS)+REDIST(IS))*
     &                 TRANRF(IS)/ASRQ(IS)

         GAG(IS)     = GTW(IS)*FAG(IS)
         GRT(IS)     = GTW(IS)*FRT(IS)
         GLV(IS)     = GAG(IS)*FLV(IS)
         GRLV(IS)    = GLV(IS)-YLV(IS)
         GRST(IS)    = GAG(IS)*FST(IS)-YST(IS)
         GRSOA(IS)   = GAG(IS)*FSOA(IS)-DSOA(IS)
         GSOB(IS)    = GAG(IS)*FSOB(IS)

      IF (IRUNLA.EQ.1) THEN
           LAID(IS) = LINT (LAITB(1,IS), ILLAI(IS), DATE)
           LAIY(IS) = LINT (LAITB(1,IS), ILLAI(IS), (DATE-1.))
           GLAI(IS) = LAID(IS)-LAIY(IS)
        ELSE
           SLA(IS)  = LINT (SLATB(1,IS), ILSLA(IS), TS(IS))
           GLAI(IS) = SLA(IS)*GRLV(IS)
             IF (LAITOT.LT.0.75) THEN
               GLAI(IS) = LAI(IS)*(EXP (RGRL(IS)*HULV(IS))-1.)
             IF (IDAY.EQ.IDAYEM(IS))
     &           GLAI(IS) = LA0(IS)*NPL(IS)*1.E-4
             ENDIF
             IF (TS(IS).GT.TSLAM(IS).AND.LAI(IS).GT.0.)
     &           GLAI(IS)= LAIM(IS)*DRL(IS)*HU(IS)
        ENDIF

        IF(WSTG(IS).GT.0.) THEN
          SSL(IS) = HGHT(IS)/((WSTG(IS)/10.)/NPL(IS))
          ELSE
          SSL(IS) = 0.
          ENDIF

        SSLMAX(IS) = EXP(AS(IS)*HGHT(IS)+BS(IS))

        IF (SSL(IS).LT.SSLMAX(IS)) THEN
          RHG(IS) = HU(IS)*HMAX(IS)*HS(IS)*HB(IS)*
     &              EXP(-HS(IS)*TS(IS))/
     &              (1.+HB(IS)*EXP(-HS(IS)*TS(IS)))**2*
     &              TRANRF(IS)
          ELSE
          RHG(IS) = 0.
          ENDIF

        ENDIF

      ENDIF

50    CONTINUE

*------- Output of states and rates only if it is required

      IF (OUTPUT .OR. TERMNL) THEN

        CALL OUTARR ('TS', TS, 1, INS)
        CALL OUTARR ('LAI', LAI, 1, INS)
        CALL OUTARR ('SAI', SAI, 1, INS)
        CALL OUTARR ('FAI', FAI, 1, INS)
```

```
          CALL OUTARR ('LAIM', LAIM, 1, INS)
          CALL OUTARR ('HGHT', HGHT, 1, INS)
          CALL OUTARR ('AMAX', AMAX, 1, INS)
          CALL OUTARR ('AMAXF', AMAXF, 1, INS)
          CALL OUTARR ('AMAXS', AMAXS, 1, INS)
          CALL OUTARR ('DTGA', DTGA, 1, INS)
          CALL OUTARR ('ASRQ', ASRQ, 1, INS)
          CALL OUTARR ('MAINT', MAINT, 1, INS)
          CALL OUTARR ('GTW', GTW, 1, INS)
          CALL OUTARR ('WDTOT', WDTOT, 1,INS)
          CALL OUTARR ('WLVG', WLVG, 1, INS)
          CALL OUTARR ('WLVD', WLVD, 1, INS)
          CALL OUTARR ('WRT', WRT, 1, INS)
          CALL OUTARR ('WSTG', WSTG, 1, INS)
          CALL OUTARR ('WSTD', WSTD, 1, INS)
          CALL OUTARR ('WSOA', WSOA, 1, INS)
          CALL OUTARR ('WSOB', WSOB, 1, INS)
          CALL OUTARR ('WSOD', WSOD, 1, INS)
          CALL OUTARR ('FRABS', FRABS, 1, INS)
          CALL OUTDAT (2,0,'ES0', ES0)
          CALL OUTDAT (2,0,'ET0', ET0)


      END IF


*=========================================================================*
*                         Integration section                            *
*=========================================================================*

      ELSE IF (ITASK.EQ.3) THEN

          LAITOT = 0.
          DO 60 IS=1,INS

*----------- Initialization of states on date of emergence

          IF (IDAY.EQ.IDAYEM(IS)) THEN
             WLVG(IS) = WLVGI(IS) * NPL(IS) * 10.
             WSTG(IS) = WSTGI(IS) * NPL(IS) * 10.
             WRT(IS)  = WRTI(IS)  * NPL(IS) * 10.
             HGHT(IS) = HGHTI(IS)
          END IF

          IF (IDAY.GT.IDAYEM(IS)) THEN
             WLVG(IS) = INTGRL (WLVG(IS), GRLV(IS) ,DELT)
             WLVG(IS) = MAX (0.,WLVG(IS))
             WLVD(IS) = INTGRL (WLVD(IS), DLV(IS)  ,DELT)
             WSTG(IS) = INTGRL (WSTG(IS), GRST(IS) ,DELT)
             WSTD(IS) = INTGRL (WSTD(IS), DST(IS)  ,DELT)
             WSOA(IS) = INTGRL (WSOA(IS), GRSOA(IS),DELT)
             WSOD(IS) = INTGRL (WSOD(IS), DSOA(IS) ,DELT)
             WRT(IS)  = INTGRL (WRT(IS) , GRT(IS)  ,DELT)
             WSOB(IS) = INTGRL (WSOB(IS), GSOB(IS) ,DELT)
             TS(IS)   = INTGRL (TS(IS)  , HU(IS)   ,DELT)
             TSLV(IS) = INTGRL (TSLV(IS), HULV(IS) ,DELT)
             HGHT(IS) = INTGRL (HGHT(IS), RHG(IS)  ,DELT)
             LAI(IS)  = INTGRL (LAI(IS) , GLAI(IS) ,DELT)
             LAI(IS)  = MAX(0.,LAI(IS))
             LAITOT   = LAITOT + LAI(IS)
                IF (TS(IS).LT.TSLAM(IS)) LAIM(IS) = LAI(IS)
```

```
      IF (TS(IS).LT.TSLAM(IS)) WLVGM(IS) = WLVG(IS)

   SAI(IS)   = SSA(IS) * WSTG(IS)
   FAI(IS)   = SFA(IS) * WSOA(IS)
   WAG(IS)   = WLVG(IS)+ WSTG(IS) + WSOA(IS)
   WTOT(IS)  = WAG(IS) + WSOB(IS)
   WDTOT(IS) = WTOT(IS)+ WLVD(IS) + WSTD(IS) + WSOD(IS)

      ENDIF

60    CONTINUE

*     Determine the finish conditions of the simulation
*     Example:
*     IF (DVS.GE.2.0)    TERMNL = .TRUE.


*=========================================================================*
*                       Terminal section                                  *
*=========================================================================*

      ELSE IF (ITASK.EQ.4) THEN

*------- Define graph for output
*       CALL OUTPLT (1,'WDTOT(1)')
*------- Use common scale, small plot width for output
*       CALL OUTPLT (7, 'INTERCOM simulation model')

      END IF

      ITOLD = ITASK

      RETURN
      END
```

Listing A2.2. Listing of the subroutine ASTRO used in the model INTERCOM.

```
*=======================================================================*
*=======================================================================*
*                                                                       *
*    SUBROUTINE ASTRO                                                    *
*    Date   : March 1992                                                 *
*    Version: 1.0                                                        *
*    Purpose: This subroutine computes daylength (DAYL) and photoperiodic *
*             daylength (DAYLP) from daynumber and latitude.             *
*                                                                       *
*                                                                       *
*    FORMAL PARAMETERS: (I=input, O=output, C=control, IN=init, T=time) *
*    name   type description                             units  class   *
*    ----   ---- -----------                             -----  -----   *
*    DAY    R4   daynumber since 1 January                 -     T,I    *
*    LAT    R4   latitude of weather station            degrees   I     *
*    DAYL   R4   daylength                                h/d    T,O    *
*    DAYLP  R4   photoperiodic daylength                  h/d    T,O    *
*    SINLD  R4   intermediate variable in calculating     -      I      *
*                daylength                                               *
*    COSLD  R4   intermediate variable in calculating     -      I      *
*                daylength                                               *
*                                                                       *
*    FATAL ERROR CHECKS (execution terminated, message): none          *
*                                                                       *
*    SUBROUTINES and FUNCTIONS called: none                            *
*                                                                       *
*    FILE usage: none                                                   *
*                                                                       *
*=======================================================================*
*=======================================================================*

      SUBROUTINE ASTRO (DAY ,LAT,
     &                  DAYL,DAYLP,SINLD,COSLD)

      IMPLICIT REAL (A-Z)
      SAVE

*---- Conversion factor from degrees to radians

      PARAMETER (PI =3.1415926)
      PARAMETER (RAD=0.017453292)

*---- Declination of the sun as function of daynumber (DAY)

      DEC  = -ASIN(SIN(23.45*RAD)*COS(2.*PI*(DAY+10.)/365.))

*---- SINLD, COSLD and AOB are intermediate variables

      SINLD = SIN(RAD*LAT)*SIN(DEC)
      COSLD = COS(RAD*LAT)*COS(DEC)
      AOB   = SINLD/COSLD

*---- Daylength (DAYL) and photoperiodic daylength (DAYLP)

      DAYL  = 12.0*(1.+2.*ASIN(AOB)/PI)
      DAYLP = 12.0*(1.+2.*ASIN((-SIN(-4.*RAD)+SINLD)/COSLD)/PI)

      RETURN
      END
```

Listing A2.3. Listing of the subroutine ASSIMC used in the model INTERCOM.

```
*======================================================================*
*======================================================================*
*                                                                      *
* SUBROUTINE ASSIMC                                                    *
* Date   : March 1992                                                  *
* Version: 1.0                                                         *
* Purpose: This subroutine (for two or more species in competition)    *
*          performs a Gaussian integration over the depth of the       *
*          canopy for each species; selects five different points      *
*          (m above soil) and computes the leaf area index above each  *
*          point (LAIC), and the leaf area density (LD) and local      *
*          assimilation rate at each point. The integrated variables   *
*          are FGROS and RADABS.                                       *
*                                                                      *
*                                                                      *
* FORMAL PARAMETERS: (I=input, O=output, C=control, IN=init, T=time)   *
* name   type description                              units  class *
* ----   ---- -----------                              -----  ----- *
* AMAX   R4   actual maximum CO2-assimilation rate      kg/ha/h   I  *
*             for individual leaves                                    *
* AMAXS  R4   actual maximum CO2-assimilation rate      kg/ha/h   I  *
*             for individual stems                                     *
* AMAXF  R4   actual maximum CO2-assimilation rate      kg/ha/h   I  *
*             for individual flowers                                   *
* INS    I4   number of species                            -      I  *
* HGHT   R4   total height of a species in the canopy      cm     I  *
* EFF    R4   initial light use efficiency for      kg/ha/h/J m2 s IN *
*             leaves                                                   *
* KDF    R4   extinction coefficient for leaves            -      I  *
* KS     R4   extinction coefficient for stems            -      I  *
* KF     R4   extinction coefficient for flowers          -      I  *
* LAI    R4   leaf area index                           ha/ha    I  *
* SAI    R4   stem area index                           m2/m2    I  *
* FAI    R4   flower area index                         m2/m2    I  *
* SINB   R4   sine of solar elevation                   m2/m2    I  *
* RADDIR R4   incoming global direct radiation          J/m/s    I  *
* RADDIF R4   incoming global diffuse radiation         J/m/s    I  *
* FGROS  R4   canopy assimilation                      kg/ha/h   O  *
* RADABS R4   absorbed radiation by species in canopy   J/m/s    O  *
*                                                                      *
* FATAL ERROR CHECKS (execution terminated, message): none            *
*                                                                      *
* SUBROUTINES and FUNCTIONS called: LEAFPA                             *
*                                                                      *
* FILE usage: none                                                     *
*                                                                      *
*======================================================================*
*======================================================================*

        SUBROUTINE ASSIMC (AMAX,AMAXS,AMAXF,INS,HGHT,EFF,KDF,KS,KF,
     &                     LAI,SAI,FAI,SINB,RADDIR,RADDIF,
     &                     FGROS,RADABS)

        IMPLICIT REAL (A-Z)
        INTEGER I,INS,IMAX,K,IG1,IG2,IGP1,IGP2
```

```
      PARAMETER (SCV=0.2, IMAX=20)
      PARAMETER (IGP1=5,  IGP2=3)
      REAL XGAUS1(IGP1),WGAUS1(IGP1), XGAUS2(IGP2), WGAUS2(IGP2)
      REAL HGHT(IMAX),  EFF(INS)
      REAL AMAX(INS),    AMAXS(INS),    AMAXF(INS)
      REAL LAI(INS),     SAI(INS),     FAI(INS)
      REAL KDF(INS),     KS(INS),      KF(INS)
      REAL FGROS(INS),   RADABS(INS)
      REAL FGL(IMAX),   AFT(IMAX)
      REAL LAIC(IMAX),  SAIC(IMAX),   FAIC(IMAX)
      REAL LD(IMAX),    SD(IMAX),     FD(IMAX)
      REAL KDFV(IMAX),  KDFSV(IMAX),  KDFFV(IMAX)
      REAL KDRBLV(IMAX),KDRBSV(IMAX), KDRBFV(IMAX)
      REAL KDRTV(IMAX) ,KDRTSV(IMAX), KDRTFV(IMAX)
      REAL FGRSH(IMAX), FGRSHS(IMAX), FGRSHF(IMAX)
      REAL FGRSUN(IMAX),FGRSUS(IMAX), FGRSUF(IMAX)
      REAL FGRS(IMAX),  VISSUN(IMAX)
      REAL AFVV(IMAX),  AFVT(IMAX),   AFVD(IMAX),   AFVSHD(IMAX)
      REAL AFVVS(IMAX), AFVTS(IMAX),  AFVDS(IMAX),  AFVSHS(IMAX)
      REAL AFVVF(IMAX), AFVTF(IMAX),  AFVDF(IMAX),  AFVSHF(IMAX)
      REAL ABSNL(IMAX), ABSNS(IMAX),  ABSNF(IMAX)

      SAVE

      DATA XGAUS1 /0.0469101,0.2307534,0.5,0.7692465,0.9530899/
      DATA WGAUS1 /0.1184635,0.2393144,0.2844444,
     &             0.2393144,0.1184635/
      DATA XGAUS2 /0.1127, 0.5000, 0.8873/
      DATA WGAUS2 /0.2778, 0.4444, 0.2778/

      PARDIF = 0.5 * RADDIF
      PARDIR = 0.5 * RADDIR

*---- Reflection coefficients of canopy for horizontal (REFVH) and
*     spherical (REFVS) leaves
      REFVH = (1.-SQRT(1.-SCV))/(1.+SQRT(1.-SCV))
      REFVS = REFVH*2./(1.+1.6*SINB)

*---- Extinction coefficients for direct component of direct flux
*     (KDIRBL), total direct flux (KDIRT), and diffuse flux (KDIF)
*     for each species; canopy assimilation is set to zero

      DO 10 K = 1,INS

*--   Leaves
      KDFV(K)  = KDF(K)
      KDRBLV(K)= (0.5/SINB)*KDFV(K)/(0.8*SQRT(1.-SCV))
      KDRTV(K) = KDRBLV(K)*SQRT(1.-SCV)

*--   Stems
      KDFSV(K) = KS(K)
      KDRBSV(K)= (0.5/SINB)*KDFSV(K)/(0.8*SQRT(1.-SCV))
      KDRTSV(K)= KDRBSV(K)*SQRT(1.-SCV)

*--   Flowers
      KDFFV(K) = KF(K)
      KDRBFV(K)= (0.5/SINB)*KDFFV(K)/(0.8*SQRT(1.-SCV))
      KDRTFV(K)= KDRBFV(K)*SQRT(1.-SCV)
```

```
        FGROS(K) = 0.
        RADABS(K)= 0.

10      CONTINUE

*---- Height (m) within canopy is selected; leaf area index
*       (LAIC, m2/m2) above each Gaussian point and leaf area
*       density (LD, m2/m3) at each point are calculated in
*       subroutine LEAFPA (or LEAFRE)

        DO 100 K  =1,INS
        DO 50  IG1=1,IGP1
        X      = XGAUS1(IG1) * HGHT(K)
        EXSDFV = 0.
        EXSTV  = 0.
        EXDV   = 0.

        DO 20 I=1,INS

        CALL LEAFPA(HGHT(I),X,LAI(I),LAIC(I),LD(I))
        CALL LEAFPA(HGHT(I),X,SAI(I),SAIC(I),SD(I))
        CALL LEAFPA(HGHT(I),X,FAI(I),FAIC(I),FD(I))

*---- Exponents for light distribution functions: sum of leaf
*       area indices above point X weighted by the extinction
*       coefficients for each species

        EXSDFV = EXSDFV + KDFV(I)    * LAIC(I)
     &                  + KDFSV(I)   * SAIC(I) + KDFFV(I)   * FAIC(I)
        EXSTV  = EXSTV  + KDRTV(I)   * LAIC(I)
     &                  + KDRTSV(I)  * SAIC(I) + KDRTFV(I)  * FAIC(I)
        EXDV   = EXDV   + KDRBLV(I)  * LAIC(I)
     &                  + KDRBSV(I)  * SAIC(I) + KDRBFV(I)  * FAIC(I)

20      CONTINUE

*---- Absorbed fluxes (J/m2 leaf/s) per species at specified
*       height in the canopy: diffuse flux, total direct flux,
*       direct component of direct flux

*--   Leaves
        AFVV(K)  = (1.-REFVH)*PARDIF*KDFV(K)   *EXP(-EXSDFV)
        AFVT(K)  = (1.-REFVS)*PARDIR*KDRTV(K)  *EXP(-EXSTV)
        AFVD(K)  = (1.-SCV)   *PARDIR*KDRBLV(K)*EXP(-EXDV)

*--   Stems
        AFVVS(K) = (1.-REFVH)*PARDIF*KDFSV(K)  *EXP(-EXSDFV)
        AFVTS(K) = (1.-REFVS)*PARDIR*KDRTSV(K) *EXP(-EXSTV)
        AFVDS(K) = (1.-SCV)   *PARDIR*KDRBSV(K)*EXP(-EXDV)

*--   Flowers
        AFVVF(K) = (1.-REFVH)*PARDIF*KDFFV(K)  *EXP(-EXSDFV)
        AFVTF(K) = (1.-REFVS)*PARDIR*KDRTFV(K) *EXP(-EXSTV)
        AFVDF(K) = (1.-SCV)   *PARDIR*KDRBFV(K)*EXP(-EXDV)

*---- Total absorbed diffuse flux (J/m2 leaf/s) and rate of
*       photosynthesis (kg/ha/h)
```

```
*--    Shaded leaves
       AFVSHD(K) = AFVV(K)+AFVT(K)-AFVD(K)
       FGRSH(K)  = AMAX(K)*(1.-EXP(-AFVSHD(K)*EFF(K)/AMAX(K)))

*--    Shaded stems
       AFVSHS(K) = AFVVS(K)+AFVTS(K)-AFVDS(K)

       IF (AMAXS(K).GT.0.) THEN
       FGRSHS(K) = AMAXS(K)*(1.-EXP(-AFVSHS(K)*EFF(K)/AMAXS(K)))
       ELSE
         FGRSHS(K) = 0.
       ENDIF

*--    Shaded flowers
       AFVSHF(K) = AFVVF(K)+AFVTF(K)-AFVDF(K)

       IF (AMAXF(K).GT.0.) THEN
       FGRSHF(K) = AMAXF(K)*(1.-EXP(-AFVSHF(K)*EFF(K)/AMAXF(K)))
       ELSE
         FGRSHF(K) = 0.
       ENDIF

*----  Direct flux absorbed by sunlit leaves perpendicular to the
*      direct beam (VISPP); instantaneous assimilation of sunlit
*      leaf area (FGRSUN) integrated over the sine of incidence of
*      direct light, assuming a spherical leaf angle distribution

*--    Leaves
       AFVPP = (1.-SCV)*PARDIR/SINB

       FGRSUN(K) = 0.
       ABSNL(K)  = 0.
       DO 30 IG2 = 1,IGP2
           VISSUN(K) = AFVSHD(K) + AFVPP      * XGAUS2(IG2)
           FGRS(K)   = AMAX(K)*(1.-EXP(-VISSUN(K)*EFF(K)/AMAX(K)))
           FGRSUN(K) = FGRSUN(K) + FGRS(K)    * WGAUS2(IG2)
           ABSNL(K)  = ABSNL(K)  + VISSUN(K)  * WGAUS2(IG2)
30     CONTINUE

*--    Stems
       FGRSUS(K) = 0.
       ABSNS(K)  = 0.

       IF (AMAXS(K).GT.0.) THEN
       DO 31 IG2 = 1,IGP2
           VISSUN(K) = AFVSHS(K) + AFVPP      * XGAUS2(IG2)
           FGRS(K)   = AMAXS(K)*(1.-EXP(-VISSUN(K)*EFF(K)/AMAXS(K)))
           FGRSUS(K) = FGRSUS(K) + FGRS(K)    * WGAUS2(IG2)
           ABSNS(K)  = ABSNS(K)  + VISSUN(K)  * WGAUS2(IG2)
31     CONTINUE
       ELSE
         FGRSUS(K) = 0.
       ENDIF

*--    Flowers
       FGRSUF(K) = 0.
       ABSNF(K)  = 0.

       IF (AMAXF(K).GT.0.) THEN
```

```
        DO 32 IG2 = 1,IGP2
           VISSUN(K) = AFVSHF(K) + AFVPP      * XGAUS2(IG2)
           FGRS(K)   = AMAXF(K)*(1.-EXP(-VISSUN(K)*EFF(K)/AMAXF(K)))
           FGRSUF(K) = FGRSUF(K) + FGRS(K)    * WGAUS2(IG2)
           ABSNF(K)  = ABSNF(K)  + VISSUN(K) * WGAUS2(IG2)
32      CONTINUE
        ELSE
         FGRSUF(K) = 0.
        ENDIF

*---- Fraction sunlit leaf area (FSLLA) and local
*     assimilation rate (FGL, kg CO2/ha leaf/h)
        FSLLAV = EXP(-EXDV)
        FGL(K) =  (FSLLAV*FGRSUN(K)+(1.-FSLLAV)*FGRSH(K)) *LD(K)
     &          +(FSLLAV*FGRSUS(K)+(1.-FSLLAV)*FGRSHS(K))*SD(K)
     &          +(FSLLAV*FGRSUF(K)+(1.-FSLLAV)*FGRSHF(K))*FD(K)

*---- Integration of local assimilation rate to canopy
*     assimilation (FGROS)
        FGROS(K) = FGROS(K) + FGL(K) * WGAUS1(IG1) * HGHT(K)

        AFT(K) =  (FSLLAV*ABSNL(K)+(1.-FSLLAV)*AFVSHD(K))*LD(K)
     &          +(FSLLAV*ABSNS(K)+(1.-FSLLAV)*AFVSHS(K))*SD(K)
     &          +(FSLLAV*ABSNF(K)+(1.-FSLLAV)*AFVSHF(K))*FD(K)

        RADABS(K) = RADABS(K) + AFT(K) * WGAUS1(IG1) * HGHT(K)

50      CONTINUE
100     CONTINUE

        RETURN
        END
```

Listing A2.4. Listing of the subroutine DEVAP used in the model INTERCOM.

```
*=======================================================================*
*=======================================================================*
*                                                                       *
* SUBROUTINE DEVAP                                                       *
* Date   : March 1992                                                   *
* Version: 1.0                                                          *
* Purpose: Computation of daily soil evaporation (AEVAP).               *
*                                                                       *
*                                                                       *
* FORMAL PARAMETERS: (I=input, O=output, C=control, IN=init, T=time)    *
* name   type description                               units  class  *
* ----   ---- -----------                               -----  ----- *
* VSMT   R4   actual soil moisture content of top layer  m3/m3    I   *
* VSMFC  R4   soil moisture content at field capacity    m3/m3    I   *
* VSMAD  R4   soil moisture content at air dryness       m3/m3    I   *
* ES0    R4   Penman evaporation of a bare soil          mm/d     I   *
* FRD    R4   fraction global radiation used for drying    -      I   *
*             power in Penman evaporation                             *
* AEVAP  R4   actual soil evaporation rate               mm/d     O   *
*                                                                       *
* FATAL ERROR CHECKS (execution terminated, message): none            *
*                                                                       *
* SUBROUTINES and FUNCTIONS called: LINT                               *
*                                                                       *
* FILE usage: none                                                      *
*                                                                       *
*=======================================================================*
*=======================================================================*

      SUBROUTINE DEVAP (VSMT,VSMFC,VSMAD,ES0,FRD,
     &                  AEVAP)

      IMPLICIT REAL (A-Z)
      REAL REDFST(12)
      SAVE
      DATA REDFST /-0.01,0., 0.2,0.05, 0.25,0.275,
     &              0.35,0.9, 1.,1., 1.1,1./

      WCPR  = (VSMT-VSMAD)/(VSMFC-VSMAD)
      REDFS = LINT(REDFST,12,WCPR)
      AEVAP = ES0*FRD*REDFS

      RETURN
      END
```

Listing A2.5. Listing of the subroutine LEAFPA used in the model INTERCOM.

```
*=====================================================================*
*=====================================================================*
*                                                                     *
* SUBROUTINE LEAFPA                                                   *
* Date   : March 1992                                                 *
* Version: 1.0                                                        *
* Purpose: This subroutine assumes a parabolic leaf area distribution; *
*          height (HGHT), a point X and total leaf area index         *
*          (LAI) are input; leaf area index of the canopy above       *
*          point X (LAIC) and the leaf area density (LD) at point X    *
*          are calculated.                                            *
*                                                                     *
*                                                                     *
* FORMAL PARAMETERS: (I=input, O=output, C=control, IN=init, T=time)  *
* name    type description                                units class *
* ----    ---- -----------                                ----- ----- *
* HGHT    R4   total height of a species in the canopy    cm    I     *
* X       R4   selected height at point X                 cm    I     *
* LAI     R4   total leaf area index                      ha/ha I     *
* LAIC    R4   total leaf area index above point X        ha/ha O     *
* LD      R4   leaf area density at point X               m2/m3 O     *
*                                                                     *
* FATAL ERROR CHECKS (execution terminated, message):                *
* X < 0.                                                              *
* HGHT < 0.                                                           *
*                                                                     *
* SUBROUTINES and FUNCTIONS called: ERROR                             *
*                                                                     *
* FILE usage: none                                                    *
*                                                                     *
*=====================================================================*
*=====================================================================*

      SUBROUTINE LEAFPA (HGHT,X,LAI,
     &                   LAIC,LD)

      IMPLICIT REAL (A-Z)
      SAVE

      IF (X .LT. 0. .OR. HGHT .LT. 0.)
     &   CALL ERROR ('LEAFPA', 'ERROR IN LEAFPA,
     &                        NEGATIVE X OR HEIGHT')

      IF (X .LE. HGHT .AND. HGHT .GT. 0.) THEN
         LAIC = LAI - ((LAI/HGHT**3) * X**2 * (3*HGHT - 2*X))
         LD   = (LAI*6./HGHT**3) * X * (HGHT - X)
      ELSE
         LAIC = 0.
         LD   = 0.
      ENDIF

      RETURN
      END
```

Listing A2.6. Listing of the subroutine LEAFRE used in the model INTERCOM.

```
*=======================================================================*
*=======================================================================*
*                                                                       *
* SUBROUTINE LEAFRE                                                     *
* Date   : March 1992                                                   *
* Version: 1.0                                                          *
* Purpose: This subroutine assumes a rectangular leaf area              *
*          distribution; height (HGHT), a point X and total leaf area   *
*          index (LAI) are input; leaf area index of the canopy above   *
*          point X (LAIC) and the leaf area density (LD) at point X     *
*          are calculated.                                              *
*                                                                       *
*                                                                       *
* FORMAL PARAMETERS: (I=input, O=output, C=control, IN=init, T=time)    *
* name    type description                               units  class  *
* ----    ---- -----------                               -----  ----- *
* HGHT    R4   total height of a species in the canopy    cm     I     *
* X       R4   selected height at point X                 cm     I     *
* LAI     R4   total leaf area index                      ha/ha  I     *
* LAIC    R4   total leaf area index above point X        ha/ha  O     *
* LD      R4   leaf area density at point X               m2/m3  O     *
*                                                                       *
* FATAL ERROR CHECKS (execution terminated, message):                   *
* X < 0.                                                                 *
* HGHT < 0.                                                             *
*                                                                       *
* SUBROUTINES and FUNCTIONS called: ERROR                              *
*                                                                       *
* FILE usage: none                                                      *
*                                                                       *
*=======================================================================*
*=======================================================================*

      SUBROUTINE LEAFRE (HGHT,X,LAI,
     &                   LAIC,LD)

      IMPLICIT REAL (A-Z)
      SAVE

      IF (X .LT. 0. .OR. HGHT .LT. 0.)
     &    CALL ERROR ('LEAFRE', 'ERROR IN LEAFRE,
     &                           NEGATIVE X OR HEIGHT')

      IF (X .LE. HGHT .AND. HGHT .GT. 0.) THEN
         LAIC = LAI * (HGHT - X)/HGHT
         LD   = LAI/HGHT
      ELSE
         LAIC = 0.
         LD   = 0.
      ENDIF

      RETURN
      END
```

Listing A2.7. Listing of the subroutine PENMAN used in the model INTERCOM.

```
*=====================================================================*
*=====================================================================*
*                                                                     *
* SUBROUTINE PENMAN                                                   *
* Date   : March 1992                                                 *
* Version: 1.0                                                        *
* Purpose: This subroutine calculates potential evapo(transpi)ration  *
*          from a free water surface (E0), a bare soil surface (ES0), *
*          and a crop (ET0) in cm/d. Input requirements are: daynumber,*
*          latitude, elevation in m, the coefficients a and b from the *
*          Angstrom formula, minimum and maximum temperature in degrees*
*          Celsius, daily total incoming radiation in J/m2, average   *
*          daily wind speed in m/s and vapour pressure in mbar.       *
*                                                                     *
*                                                                     *
* FORMAL PARAMETERS: (I=input, O=output, C=control, IN=init, T=time)  *
* name    type description                           units  class *
* ----    ---- -----------                           -----  ----- *
* TMN     R4   daily minimum air temperature         degrees Celsius I *
* TMX     R4   daily maximum air temperature         degrees Celsius I *
* AVRAD   R4   daily incoming total global radiation    J/m2/d   I *
* ATMTR   R4   atmospheric transmission coefficient       -      I *
* WIND    R4   daily average wind speed                  m/s     I *
* VAPOUR  R4   average vapour pressure                   mbar    I *
* E0      R4   Penman evaporation of a free water surface  mm/d  O *
* ES0     R4   Penman evaporation of a bare soil          mm/d   O *
* ET0     R4   Penman evaporation of a crop stand         mm/d   O *
*                                                                     *
* FATAL ERROR CHECKS (execution terminated, message): none           *
*                                                                     *
* SUBROUTINES and FUNCTIONS called: ERROR                            *
*                                                                     *
* FILE usage: none                                                    *
*                                                                     *
*=====================================================================*
*=====================================================================*

      SUBROUTINE PENMAN (TMN,TMX,AVRAD,ATMTR,WIND,VAPOUR,
     &                   E0,ES0,ET0)

      IMPLICIT REAL (A-Z)
      SAVE

*---- Albedo, reflection-coefficient, for water surface (REFCFW),
*     soil surface (REFCFS) and canopy (REFCFC).
*     PAR: reflection=.10, transmission=.10, absorption=.80
*     NIR: reflection=.40, transmission=.40, absorption=.20
*     AVRAD consists of 50% PAR and 50% NIR, so
*     REFCFC = (.10 + .40)/2 = .25
      DATA REFCFW /0.05/
      DATA REFCFS /0.15/
      DATA REFCFC /0.25/

*---- Latent heat of evaporation of water (J/kg=J/mm) and
*     Stefan-Boltzmann constant (J/m2/d/K), psychrometric
*     instrument constant (K-1)
```

```
        DATA LHVAP /2.45E6/
        DATA STBC  /4.9E-3/
        DATA PSYCON/0.000662/
        DATA A /0.18/, B /0.55/

*---- Mean daily temperature and temperature difference (Celsius)
        TMPA = (TMN+TMX)/2.
        TDIF = TMX-TMN

*---- Coefficient Bu in wind function, dependent on
*       temperature difference
        BU = 0.54+0.35*LIMIT(0.,1.,(TDIF-12.)/4.)

*---- Barometric pressure (mbar), psychrometric constant (mbar/K)
*       GAMMA is dependent on sea level (ELEV).
        PBAR  = 1013.
        GAMMA = PSYCON*PBAR

*---- Saturated vapour pressure according to equation
*       of Goudriaan (1977)
        SVPA = 6.11*EXP(17.4*TMPA/(TMPA+239.))

        IF (VAPOUR.GT.SVPA*1.1) CALL ERROR ('PENMAN',
      &    'Actual vapour pressure greater than saturated')

*---- Derivative of SVPA with respect to temperature, i.e. slope
*       of the SVPA-temperature curve (mbar/K)
        DELTA = 239.*17.4*SVPA/(TMPA+239.)**2

*---- The expression n/N (RELSSD) from the Penman formula is
*       estimated from the Angstrom formula: RI=RA(A+B.n/N) ->
*       n/N=(RI/RA-A)/B, where RI=AVRAD (daily total global
*       radiation) and RA=ANGOT (total daily radiation with
*       a clear sky, the Angot radiation), n = hours of sunshine,
*       N = daylength in hours. A = 0.25, B = 0.75 according to
*       de Wit et al., 1978, these values are general estimates;
*       for the UK, 52 NL:  A=0.18 and B=0.55
        RELSSD = LIMIT (0.,1.,(ATMTR-A)/B)

*---- Terms of the Penman formula, for water surface, soil
*       surface and canopy; net outgoing long-wave radiation
*       (J/m2/d) according to Brunt (1932)
        RB = STBC*(TMPA+273.)**4*(0.56-0.079*SQRT(VAPOUR))*
      &      (0.1+0.9*RELSSD)

*---- Net absorbed radiation, expressed in mm/d
        RNW = (AVRAD*(1.-REFCFW)-RB)/LHVAP
        RNS = (AVRAD*(1.-REFCFS)-RB)/LHVAP
        RNC = (AVRAD*(1.-REFCFC)-RB)/LHVAP

*---- Evaporative demand of the atmosphere (mm/d)
        EA  = 0.26*(SVPA-VAPOUR)*(0.5+BU*WIND)
        EAC = 0.26*(SVPA-VAPOUR)*(1.0+BU*WIND)

*---- Penman formula (1948), and conversion to mm/d.
        E0      = (DELTA*RNW+GAMMA*EA)/(DELTA+GAMMA)
        ES0     = (DELTA*RNS+GAMMA*EA)/(DELTA+GAMMA)
        ET0     = (DELTA*RNC+GAMMA*EAC)/(DELTA+GAMMA)

        RETURN
        END
```

Listing A2.8. Listing of the subroutine RADIAT used in the model INTERCOM.

```
*=====================================================================*
*=====================================================================*
*                                                                     *
* SUBROUTINE RADIAT                                                   *
* Date   : March 1992                                                *
* Version: 1.0                                                       *
* Purpose: This subroutine computes diffuse and direct amount of     *
*          photosynthetically active radiation from average global   *
*          radiation (AVRAD), day of the year and hour of the day.   *
*                                                                     *
*                                                                     *
* FORMAL PARAMETERS: (I=input, O=output, C=control, IN=init, T=time)  *
* name    type description                               units  class *
* ----    ---- -----------                               -----  ----- *
* HOUR    R4   selected hour at which CO2 assimilation      h     T,I  *
*              is calculated                                          *
* DAY     R4   daynumber since 1 January                    d     T,I  *
* DAYL    R4   daylength                                   h/d    T,I  *
* SINLD   R4   intermediate variable                        -      I   *
* COSLD   R4   intermediate variable                        -      I   *
* AVRAD   R4   daily incoming total global radiation      J/m2/d   I   *
* ATMTR   R4   atmospheric transmission coefficient         -      I   *
* SINB    R4   sine of solar elevation                      _      I   *
* PARDIR  R4   instantaneous flux of direct PAR           J/m2/s   O   *
* PARDIF  R4   instantaneous flux of diffuse PAR          J/m2/s   O   *
*                                                                     *
* FATAL ERROR CHECKS (execution terminated, message): none           *
*                                                                     *
* SUBROUTINES and FUNCTIONS called: none                             *
*                                                                     *
* FILE usage: none                                                   *
*                                                                     *
*=====================================================================*
*=====================================================================*

      SUBROUTINE RADIAT (HOUR,DAY,DAYL,SINLD,COSLD,AVRAD,
     &                   ATMTR,SINB,PARDIR,PARDIF)

      IMPLICIT REAL (A-Z)
      PARAMETER (PI=3.1415926)

      SAVE

*---- Sine of solar elevation (SINB), integral of SINB (DSINB)
*     and integral of SINB with correction for lower atmospheric
*     transmission at low solar elevations (DSINBE)

      AOB   = SINLD/COSLD
      SINB  = AMAX1(0.,SINLD+COSLD*COS(2.*PI*(HOUR+12.)/24.))
      DSINB = 3600.*(DAYL*SINLD+24.*COSLD*SQRT(1.-AOB*AOB)/PI)
      DSINBE= 3600.*(DAYL*(SINLD+0.4*
     &              (SINLD*SINLD+COSLD*COSLD*0.5))+12.0*COSLD*
     &              (2.0+3.0*0.4*SINLD)*SQRT(1.-AOB*AOB)/PI)

*---- Solar constant (SC) and daily extraterrestrial
```

```
*      radiation (ANGOT)

       SC    = 1370.*(1.+0.033*COS(2.*PI*DAY/365.))
       ANGOT = SC * DSINB

*---- Diffuse light fraction (FRDIF) from atmospheric
*      transmission (ATMTR)

       ATMTR = AVRAD/ANGOT
       IF (ATMTR.GT.0.75) FRDIF = 0.23
       IF (ATMTR.LE.0.75.AND.ATMTR.GT.0.35)
      &                   FRDIF = 1.33-1.46*ATMTR
       IF (ATMTR.LE.0.35.AND.ATMTR.GT.0.07)
      &                   FRDIF = 1.-2.3*(ATMTR-0.07)**2
       IF (ATMTR.LE.0.07) FRDIF = 1.

*---- Diffuse PAR (PARDIF) and direct PAR (PARDIR)

       PAR    = 0.5*AVRAD*SINB*(1.+0.4*SINB)/DSINBE
       PARDIF = AMIN1(PAR,SINB*FRDIF*ATMTR*0.5*SC)
       PARDIR = PAR-PARDIF

       RETURN
       END
```

Listing A2.9. Listing of the subroutine TOTASS used in the model INTERCOM.

```
*======================================================================*
*======================================================================*
*                                                                      *
* SUBROUTINE TOTASS                                                    *
* Date    : March 1992                                                 *
* Version: 1.0                                                         *
* Purpose: This subroutine calculates daily total gross assimilation   *
*          (DTGA) by performing a Gaussian integration over time. At   *
*          three different times of the day, radiation is computed and *
*          used to determine assimilation whereafter integration       *
*          takes place.                                                *
*                                                                      *
*                                                                      *
* FORMAL PARAMETERS: (I=input, O=output, C=control, IN=init, T=time)   *
* name     type description                            units   class  *
* ----     ---- -----------                            -----   -----   *
* DAY      R4   daynumber since 1 January                d      T,I   *
* DAYL     R4   daylength                              h/d      T,I   *
* KDF      R4   extinction coefficient for leaves        -       I    *
* KS       R4   extinction coefficient for stems         -       I    *
* KF       R4   extinction coefficient for flowers       -       I    *
* AMAX     R4   actual maximum CO2-assimilation rate for kg/ha/h  I    *
*               individual leaves                                      *
* AMAXS    R4   actual maximum CO2-assimilation rate for kg/ha/h  I    *
*               individual stems                                       *
* AMAXF    R4   actual maximum CO2-assimilation rate for kg/ha/h  I    *
*               individual flowers                                     *
* EFF      R4   initial light use efficiency for     kg/ha/h/J m2 s IN *
*               leaves                                                 *
* LAI      R4   leaf area index                        ha/ha     I    *
* SAI      R4   stem area index                        m2/m2     I    *
* FAI      R4   flower area index                      m2/m2     I    *
* AVRAD    R4   daily incoming total global radiation  J/m2/d    I    *
* SINLD    R4   intermediate variable in calculating     -       I    *
*               daylength                                              *
* COSLD    R4   intermediate variable in calculating     -       I    *
*               daylength                                              *
* INS      I4   number  of species                       -       I    *
* HGHT     R4   total height of a species in the canopy  cm      I    *
* ATMTR    R4   atmospheric transmission coefficient      -       I    *
* FRABS    R4   fraction absorbed incoming global radiation -     I    *
* FRD      R4   fraction global radiation used for drying  -      I    *
*               power in Penman evaporation                            *
* DTGA     R4   daily total gross CO2-assimilation     kg/ha/h   O    *
*                                                                      *
* FATAL ERROR CHECKS (execution terminated, message): none            *
*                                                                      *
* SUBROUTINES and FUNCTIONS called: RADIAT, ASSIMC                     *
*                                                                      *
* FILE usage: none                                                     *
*                                                                      *
*======================================================================*
*======================================================================*

    SUBROUTINE TOTASS (DAY,DAYL,KDF,KS,KF,AMAX,AMAXS,AMAXF,EFF,
```

```
     &                          LAI,SAI,FAI,AVRAD,SINLD,COSLD,INS,HGHT,
     &                          ATMTR,FRABS,FRD,DTGA)

      IMPLICIT REAL (A-Z)
      INTEGER I,INS,K,INGP,IMAX
      PARAMETER (INGP=3, IMAX=20)

      REAL XGAUSS(INGP), WGAUSS(INGP)
      REAL DTGA(INS),FGROS(IMAX),FRABS(INS),KDF(INS),AMAX(INS),
     &     LAI(INS),EFF(INS),HGHT(INS),DRABS(IMAX),RADABS(IMAX)
      REAL KS(INS),KF(INS),SAI(INS),FAI(INS),AMAXS(INS),AMAXF(INS)

      SAVE
      DATA WGAUSS /0.2778, 0.4444, 0.2778/
      DATA XGAUSS /0.1127, 0.5000, 0.8873/

*---- Assimilation set to zero and three different times of
*     the day (HOUR)

      DO 10 K=1,INS
         DTGA(K)  = 0.
         DRABS(K) = 0.
10    CONTINUE

      DO 30 I = 1,INGP
         HOUR = 12.0 + DAYL * 0.5 * XGAUSS(I)

*------- At the specified HOUR, radiation is computed and used
*        to compute assimilation

         CALL RADIAT (HOUR,DAY,DAYL,SINLD,COSLD,AVRAD,
     &                ATMTR,SINB,PARDIR,PARDIF)

         RADDIF = 2. * PARDIF
         RADDIR = 2. * PARDIR

         CALL ASSIMC (AMAX,AMAXS,AMAXF,INS,HGHT,EFF,KDF,KS,KF,
     &                LAI,SAI,FAI,SINB,RADDIR,RADDIF,
     &                FGROS,RADABS)

*------- Integration of assimilation rate to a daily total (DTGA)

         DO 20 K=1,INS
            DTGA(K)  = DTGA(K) + FGROS(K)  * WGAUSS(I)
            DRABS(K) = DRABS(K)+ RADABS(K) * WGAUSS(I)
20       CONTINUE

30    CONTINUE
      LAITOT = 0.
      TDRABS = 0.
      DO 40 K=1,INS
         DTGA(K)  = DTGA(K)  * DAYL
         DRABS(K) = DRABS(K) * DAYL * 3600.
         TDRABS   = TDRABS + DRABS(K)
         LAITOT   = LAITOT + LAI(K)
40    CONTINUE

      DO 45 K=1,INS
        IF (TDRABS.GT.0.) THEN
```

```
            FRABS(K) = (DRABS(K)/TDRABS)*(1.-EXP(-0.5*LAITOT))
      ELSE
            FRABS(K) = 0.
      ENDIF
45    CONTINUE

      FRD = EXP(-0.5*LAITOT)

      RETURN
      END
```

Listing A2.10. Listing of the subroutine TOTRAN used in the model INTERCOM.

```
*========================================================================*
*========================================================================*
*                                                                        *
* SUBROUTINE TOTRAN                                                       *
* Date   : March 1992                                                    *
* Version: 1.0                                                           *
* Purpose: This subroutine calculates the factor accounting for the      *
*          of water stress on the rate of dry matter increase (TRANRF),  *
*          the actual transpiration rate of a species (TRAN) and the     *
*          total actual transpiration rate of the canopy (ATRAN).        *
*                                                                        *
*                                                                        *
*                                                                        *
* FORMAL PARAMETERS: (I=input, O=output, C=control, IN=init, T=time)     *
* name    type description                                  units  class *
* ----    ---- -----------                                  -----  ----- *
* INS     I4   number of species                              -      I   *
* TMN     R4   daily minimum temperature          degrees Celsius    I   *
* TMX     R4   daily maximum temperature          degrees Celsius    I   *
* AVRAD   R4   daily incoming total global radiation      J/m2/d     I   *
* ATMTR   R4   atmospheric transmission coefficient         -        I   *
* WIND    R4   daily average windspeed                     m/s       I   *
* VAPOUR  R4   average vapour pressure                     mbar      I   *
* CRPF    R4   crop factor for transpiration                -        I   *
* VSMFC   R4   soil moisture content at field capacity    m3/m3      I   *
* VSMRTZ  R4   actual soil moisture content root layer     m3/m3     I   *
* FRABS   R4   fraction absorbed incoming global radiation   -       I   *
* PTB     R4   table of soil depletion factor versus ET0  -,mm/d     I   *
* ILPTB   I4   number of elements of array PTB               -        I   *
* E0      R4   Penman evaporation of a free water surface  mm/d      I   *
* ES0     R4   Penman evaporation of a bare soil           mm/d      I   *
* ET0     R4   Penman evaporation of a crop stand          mm/d      I   *
* TRAN    R4   actual transpiration rate of a species      mm/d      O   *
* ATRAN   R4   total actual transpiration rate of the canopy mm/d    O   *
* TRANRF  R4   factor accounting for effect of water stress  -       O   *
*              on the rate of dry matter increase                        *
*                                                                        *
* FATAL ERROR CHECKS (execution terminated, message): none               *
*                                                                        *
* SUBROUTINES and FUNCTIONS called: PENMAN, LINT                         *
*                                                                        *
* FILE usage: none                                                       *
*                                                                        *
*========================================================================*
*========================================================================*

      SUBROUTINE TOTRAN (INS,TMN,TMX,AVRAD,ATMTR,WIND,VAPOUR,
     &                   CRPF,VSMFC,VSMWP,VSMRTZ,FRABS,PTB,ILPTB,
     &                   E0,ES0,ET0,TRAN,ATRAN,TRANRF)

      IMPLICIT REAL (A-Z)
      INTEGER INS,IS,IMNS,IMNP
      PARAMETER (IMNS=10,IMNP=40)
      INTEGER ILPTB(IMNS)
      REAL PTRAN(IMNS),FRABS(IMNS)
      REAL CRPF(IMNS), TRANRF(IMNS)
```

```
      REAL P(IMNS)
      REAL VSMCR(IMNS)
      REAL TRAN(IMNS)
      REAL PTB(IMNP,IMNS)

      SAVE

      CALL PENMAN (TMN,TMX,AVRAD,ATMTR,WIND,VAPOUR,
     &              E0,ES0,ET0)

      ATRAN = 0.

      DO 45 IS=1,INS
         PTRAN(IS)  = ET0*FRABS(IS)*CRPF(IS)
         P(IS)      = LINT (PTB(1,IS),ILPTB(IS),ET0)
         VSMCR(IS)  = (1.-P(IS))*(VSMFC-VSMWP)+VSMWP
         TRANRF(IS) = LIMIT(0.,1.,(VSMRTZ-VSMWP)/(VSMCR(IS)-VSMWP))
         TRAN(IS)   = PTRAN(IS)*TRANRF(IS)
         ATRAN      = (ATRAN+TRAN(IS))
45    CONTINUE

      RETURN
      END
```

Listing A2.11. Listing of the subroutine WBAL used in the model INTERCOM.

```
*========================================================================*
*========================================================================*
*                                                                        *
* SUBROUTINE WBAL                                                        *
* Date   : March 1992                                                   *
* Version: 1.0                                                          *
* Purpose: This subroutine is a simple water balance of the soil and    *
*          calculates the actual volumetric soil moisture content       *
*          of the top layer of the soil (VSMT) and of the root          *
*          layer (VSMRTZ).                                              *
*                                                                        *
*                                                                        *
*                                                                        *
* FORMAL PARAMETERS: (I=input, O=output, C=control, IN=init, T=time)     *
* name    type description                                units class *
* ----    ---- -----------                                ----- ----- *
* ITASK   I4   determines action of the subroutine,         -    C,I  *
*              1=initialization, 2=rate calculation,                    *
*              3=integration, 4=terminal                               *
* IUNITS  I4   unit number of soil data file                -    C,IN *
* IUNITO  I4   unit number of output file                   -    C,IN *
* FILES   C*   file name for soil variables                 -    C,IN *
* OUTPUT  L4   flag that indicates if output to file is     -    C,I  *
*              required                                                 *
* TERMNL  L4   flag that indicates if simulation should     -   C,I,O *
*              terminate                                               *
* DATE    R4   daynumber since 1 January                    -    T,I  *
* DELT    R4   time interval of integration                 -    T,I  *
* RAIN    R4   water input through rainfall               mm/d   I    *
* AEVAP   R4   actual soil evaporation rate               mm/d   I    *
* ATRAN   R4   total actual transpiration rate of the canopy mm/d I    *
* RTD     R4   rooted depth                                m     I    *
* VSMFC   R4   soil moisture content at field capacity    m3/m3  I    *
* VSMAD   R4   soil moisture content at air dryness       m3/m3  I    *
* VSMWP   R4   soil moisture content at wilting point     m3/m3  I    *
* VSMT    R4   actual soil moisture content top layer soil m3/m3 I    *
* VSMRTZ  R4   actual soil moisture content root layer    m3/m3  I    *
*                                                                        *
* FATAL ERROR CHECKS (execution terminated, message):                   *
* Certain sequences of ITASK, see subroutine CHKTSK                     *
*                                                                        *
* SUBROUTINES and FUNCTIONS called: CHKTSK, COPFIL, RDINIT, RDAREA,      *
*                                   RDSREA, LINT, OUTDAT, OUTARR,        *
*                                   INTGRL                              *
*                                                                        *
* FILE usage: soil variables file FILES with unit IUNITS                *
*                                                                        *
*========================================================================*
*========================================================================*

      SUBROUTINE WBAL (ITASK,IUNITS,IUNITO,FILES,OUTPUT,TERMNL,
     &                 DATE,DELT,
     &                 RAIN,AEVAP,ATRAN,RTD,
     &                 VSMFC,VSMAD,VSMWP,VSMT,VSMRTZ)

      IMPLICIT REAL (A-Z)
```

```
*---- Formal parameters
      INTEGER ITASK,IUNITS,IUNITO
      LOGICAL OUTPUT,TERMNL
      CHARACTER*(*) FILES

*---- Local parameters
      INTEGER ITOLD
      INTEGER IRIRR,IRCAP
      REAL RIRRIT(100), RCAPT(100)
      LOGICAL INIT
      SAVE
      DATA ITOLD /4/, INIT /.FALSE./

      CALL CHKTSK ('WBAL', IUNITO, ITOLD, ITASK)

      IF (ITOLD.EQ.1.AND.ITASK.EQ.3) THEN
         ITOLD = ITASK
         RETURN
      END IF

      IF (ITASK.EQ.1) THEN

         IF (.NOT.INIT) THEN
            CALL COPFIL (IUNITS, FILES, IUNITO)
            INIT = .TRUE.
         END IF
         CALL RDINIT (IUNITS, IUNITO, FILES)
         CALL RDAREA ('RIRRIT', RIRRIT,100,IRIRR)
         CALL RDAREA ('RCAPT', RCAPT,100,IRCAP)
         CALL RDSREA ('THCKT', THCKT)
         CALL RDSREA ('SMTI', SMTI)
         SMT = SMTI
         CALL RDSREA ('SMRTZI', SMRTZI)
         CALL RDSREA ('TCS', TCS)
         CALL RDSREA ('RTDMAX', RTDMAX)
         CALL RDSREA ('VSMFC', VSMFC)
         CALL RDSREA ('VSMWP', VSMWP)
         VSMAD = VSMWP/3.
         SMFCT = VSMFC*THCKT
         RTDT  = RTDMAX
         SMFRTZ= VSMFC*RTDT

         CLOSE (IUNITS, STATUS='DELETE')

         SMRTZ = SMRTZI
         SMRTD = SMT+SMRTZ

         TAEVAP = 0.
         TATRAN = 0.
         RAINT  = 0.
         VSMT   = SMT/THCKT
         VSMRTZ = SMRTZ/RTDT

      ELSE IF (ITASK.EQ.2) THEN

         RIRRI = LINT (RIRRIT,IRIRR,DATE)
         RCAP  = LINT (RCAPT,IRCAP,DATE)
         INFR  = RAIN+RIRRI
```

```
*---- Distribution of evaporation over layers
      AEVAPT = AEVAP*0.26
      AEVAPR = AEVAP-AEVAPT

*---- Water flow over borders of two compartments
      PERC   = MAX(0.,SMT+INFR-AEVAPT-SMFCT)
      RDRAIN = MAX(0.,SMRTZ+PERC+RCAP-AEVAPR-ATRAN-SMFRTZ)/TCS

         IF (OUTPUT.OR.TERMNL) THEN
*            CALL OUTDAT (2,0,'AEVAP', AEVAP)
*            CALL OUTARR ('TRAN', TRAN, INS)
*            CALL OUTDAT (2,0,'ATRAN', ATRAN)
*            CALL OUTARR ('TRANRF', TRANRF, INS)
*            CALL OUTDAT (2,0,'SMRTD',SMRTD)
*            CALL OUTDAT (2,0,'RAINT',RAINT)
*            CALL OUTDAT (2,0,'TAEVAP',TAEVAP)
*            CALL OUTDAT (2,0,'TATRAN',TATRAN)
         END IF

      ELSE IF(ITASK.EQ.3) THEN

      RTDT   = MIN (RTD,RTDMAX)
      SMT    = INTGRL (SMT,INFR-AEVAPT-PERC,DELT)
      SMRTZ  = INTGRL (SMRTZ,PERC+RCAP-AEVAPR-ATRAN-RDRAIN,DELT)
      TAEVAP = INTGRL (TAEVAP,AEVAP,DELT)
      TATRAN = INTGRL (TATRAN,ATRAN,DELT)
      SMRTD  = SMT+SMRTZ
      RAINT  = RAINT+RAIN
      VSMT   = SMT/THCKT
      VSMRTZ = SMRTZ/RTDT

      ELSE IF(ITASK.EQ.4) THEN

      ENDIF

      ITOLD = ITASK

      RETURN
      END
```

Listing A2.12. Listing of the plant data set for sugar beet (SUGARB.DAT) used in the model INTERCOM. For explanation of the abbrevations see Appendix 3.

```
*****************************************************************
* Plant data set for Sugar beet                                *
*****************************************************************
RTD    = 0.6
AMAXM  = 50.
AMAXS  = 10.
AMAXF  = 10.
EFFTB  = 0.,0.5, 40.,0.5
KDF    = 0.69; KS     = 0.6; KF     = 0.6
CRPF   = 1.0
PTB    = 0.,0.75,  8.,0.11
REDFTB = -10.,1.,  0.,1., 40.,1.
REDMNT = -10.,1.,  0.,1., 40.,1.
AMDVST =   0.,1.,   500.,1., 3000.,1.
TMD    = 2.;  TMDLV  = 3.; TMXD = 21.; TMXLV = 21.
MAINLV = 0.03; MAINST = 0.01
MAINSA = 0.01; MAINSB = 0.002
MAINRT = 0.01
FAGTB  =     0.,0.8, 1000.,0.9, 2000.,1. , 3000.,1.
FLVTB  =     0.,0.9,  250.,0.9,  350.,0.7,  550.,0.4, 950.,0.05,
          3000.,0.05
FSTTB  =     0.,0.,   250.,0. ,  350.,0.2,  550.,0.3, 950.,0.3 ,
          1200.,0.15, 3000.,0.15
FSOATB =   0.,0.,   3000.,0.
FRTTB  =   0.,0.2, 1000.,0.1, 2000.,0., 3000.,0.
ASRQLV = 1.46; ASRQST = 1.51
ASRQSA = 1.51; ASRQRT = 1.44
ASRQSB = 1.29
DRL    = -0.00055
TSLAM  = 1329.
RDRSTT =   0.,0.,        600.,0.,     1000.,0.00022, 1500.,0.00050,
          2500.,0.00075, 3000.,0.00075
REDLM  = 0.13
REDST  = 0.36
RDRSOT = 0.,0., 3000.,0.
*LAI table for 1985 in mixture, C. album low density
LAITB = 117.,0., 128.,0.,    129.,0.00056, 140.,0.0043,
                 154.,0.102, 168.,0.792,   182.,1.98,  203.,3.97,
                 224.,4.17,  247.,4.19,    269.,3.27
*LAI table for 1985 Sugar beet in monoculture
*LAITB = 117.,0., 128.,0.,    129.,0.00056, 140.,0.00351,
*                 154.,0.123, 168.,0.772,   182.,2.35, 203.,5.65,
*                 224.,5.10,  247.,5.10,    269.,4.66
SLATB  = 0.,0.0025, 250.,0.0020, 3000.,0.00200
SSA    = 0.0;  SFA    = 0.0
RGRL   = 0.0158
HMAX   = 60.
HS     = 0.007
HB     = 67.
* Initial states
LA0 = 0.45; WLVGI=0.; WSTGI=0.; WRTI=0.; HGHTI=0.1
*
*****************************************************************
```

Listing A2.13. Listing of the plant data set for *C. album* (CHENO.DAT) used in the model INTERCOM. For explanation of the abbrevations see Appendix 3.

```
******************************************************************
* Plant data set for Chenopodium album                          *
******************************************************************
RTD     = 0.6
AMAXM   = 50.
AMAXS   = 10.
AMAXF   = 10.
EFFTB   = 0.,0.5, 14.,0.49, 21.,0.46, 28.,0.39, 38.,0.31
KDF     = 0.69; KS      = 0.69; KF      = 0.69
CRPF    = 1.0
PTB     = 0.,0.75, 8.,0.11
REDFTB  = -10.,1., 0.,1., 40.,1.
REDMNT  = -10.,1., 0.,1., 40.,1.
AMDVST  = 0.,1., 1300.,1., 2200.,0.
TMD     = 2.;  TMDLV  = 3.;  TMXD = 21.;  TMXLV = 21.
MAINLV  = 0.03; MAINST = 0.015
MAINSA  = 0.01; MAINSB = 0.0
MAINRT  = 0.01
FAGTB   =    0.,0.84,  429.,0.84,  430.,1.  , 3000.,1.
FLVTB   =    0.,0.89,  120.,0.89,  400.,0.67, 500.,0.485, 600.,0.3,
           800.,0.17,  950.,0.  , 3000.,0.
FSTTB   =    0.,0.11,  120.,0.11,  400.,0.33, 500.,0.515, 600.,0.63,
           800.,0.63,  950.,0.3, 1250.,0.  , 3000.,0.
FSOATB  =    0.,0.  ,  500.,0.  ,  600.,0.07, 800.,0.2,   950.,0.7,
          1250.,1.  , 3000.,1.
FRTTB   =    0.,0.16,  429.,0.16,  430.,0.  , 3000.,0.
ASRQLV  = 1.56; ASRQST = 1.51
ASRQSA  = 1.49; ASRQRT = 1.44
ASRQSB  = 0.
DRL     = -0.0022
TSLAM   = 1267.
RDRSTT  = 0.,0.,  3000.,0.0
REDLM   = 0.0; REDST  = 0.0
RDRSOT  = 0.,0., 1100.,0., 1402.,0.0013,1666.,0.008, 3000.,0.1
*LAI table for 1985 mixture, low density (NPL=5.5)
LAITB   = 117.,0.,     138.,0.,     154.,0.00214, 168.,0.0397,
          182.,0.235, 202.,0.940, 223.,0.798,    244.,0.584,
          266.,0.121, 269.,0.121
*LAI table for 1985 monoculture (NPL=11)
*LAITB  = 117.,0.,     138.,0.,     154.,0.00545, 168.,0.0849,
*         182.,0.800, 202.,4.59, 223.,4.71,    244.,4.23,
*         266.,1.13, 269.,1.13
SLATB   = 0.,0.0023, 3000.,0.0023
SSA     = 0.00010; SFA    = 0.00010
RGRL    = 0.0186
HMAX    = 160.
HS      = 0.009
HB      = 298.
AS      = -0.04
BS      = 6.8
* Initial states
LA0 = 0.13; WLVGI=0.; WSTGI=0.; WRTI=0.; HGHTI=0.1
*
******************************************************************
```

Appendix Three

Definition of the Abbreviations Used in the Model INTERCOM

Name	Description	Routine	Units
A	Parameter in Ångström formula	PENMAN MAIN	-
ABSNF	Absorbed radiation by sunlit flower area at a specific height in the canopy	ASSIMC	$J\ m^{-2}$ flower s^{-1}
ABSNL	Absorbed radiation by sunlit leaf area at a specific height in the canopy	ASSIMC	$J\ m^{-2}$ leaf s^{-1}
ABSNS	Absorbed radiation by sunlit stem area at a specific height in the canopy	ASSIMC	$J\ m^{-2}$ stem s^{-1}
AEVAP	Actual soil evaporation rate, derived from Penman evaporation	DEVAP MAIN WBAL	$mm\ d^{-1}$
AEVAPR	Actual soil evaporation rate from second soil layer	WBAL	$mm\ d^{-1}$
AEVAPT	Actual soil evaporation rate from the top layer of the soil	WBAL	$mm\ d^{-1}$
AFT	Absorbed flux at selected canopy height total (PAR)	ASSIMC	$J\ m^{-2}$ ground s^{-1}
AFVD	Absorbed flux direct component of direct flux (PAR) at selected canopy height by leaves	ASSIMC	$J\ m^{-2}$ leaf s^{-1}
AFVDF	Absorbed flux direct component of direct flux (PAR) at selected canopy height by flowers	ASSIMC	$J\ m^{-2}$ flower s^{-1}
AFVDS	Absorbed flux direct component of direct flux (PAR) at selected canopy height by stems	ASSIMC	$J\ m^{-2}$ stem s^{-1}
AFVPP	Absorbed flux by sunlit area (leaves, stems, flowers) (PAR) perpendicular to direct beam	ASSIMC	$J\ m^{-2}$ tissue s^{-1}
AFVSHD	Absorbed flux (PAR) by shaded leaf area	ASSIMC	$J\ m^{-2}$ leaf s^{-1}
AFVSHF	Absorbed flux (PAR) by shaded flower area	ASSIMC	$J\ m^{-2}$ flower s^{-1}
AFVSHS	Absorbed flux (PAR) by shaded stem area	ASSIMC	$J\ m^{-2}$ stem s^{-1}
AFVT	Absorbed flux total direct (PAR) at selected canopy height by leaves	ASSIMC	$J\ m^{-2}$ leaf s^{-1}
AFVTF	Absorbed flux total direct (PAR) at selected canopy	ASSIMC	$J\ m^{-2}$ flower s^{-1}

	height by flowers		
AFVTS	Absorbed flux total direct (PAR) at selected canopy height by stems	ASSIMC	$J\ m^{-2}\ stem\ s^{-1}$
AFVV	Absorbed diffuse flux (PAR) at selected canopy height by leaves	ASSIMC	$J\ m^{-2}\ leaf\ s^{-1}$
AFVVF	Absorbed diffuse flux (PAR) at selected canopy height by flowers	ASSIMC	$m^{-2}\ flower\ s^{-1}$
AFVVS	Absorbed diffuse flux (PAR) at selected canopy height by stems	ASSIMC	$J\ m^{-2}\ stem\ s^{-1}$
AMAX	Actual maximum CO_2 assimilation rate at light saturation for individual leaves	ASSIMC PLANTC TOTASS	$kg\ CO_2$ $ha^{-1}\ leaf\ h^{-1}$
AMAXF	Actual maximum CO_2 assimilation rate at light saturation for individual flowers	ASSIMC PLANTC TOTASS	$kg\ CO_2$ $ha^{-1}\ flower\ h^{-1}$
AMAXM	Potential maximum CO_2 assimilation rate at light saturation for individual leaves	PLANTC	$kg\ CO_2$ $ha^{-1}\ leaf\ h^{-1}$
AMAXS	Actual maximum CO_2 assimilation rate at light saturation for individual stems	ASSIMC PLANTC TOTASS	$kg\ CO_2$ $ha^{-1}\ stem\ h^{-1}$
AMDVS	Reduction factor accounting for effect of development stage on AMAXM	PLANTC	-
AMDVST	Table of AMDVS as a function of temperature sum (or DVS)	PLANTC	-, $^{\circ}C\ d$, (-)
ANGOT	Daily extra-terrestrial radiation	RADIAT	$J\ m^{-2}\ d^{-1}$
AOB	Intermediate variable in calculating daylength and solar sine	ASTRO RADIAT	-
AS	Parameter for calculation SSLMAX	PLANTC	m^{-1}
ASRQ	Assimilate requirement for total plant dry matter production	PLANTC	$kg\ CH_2O$ $kg^{-1}\ DM\ crop$
ASRQLV	Assimilate requirement for leaf dry matter production	PLANTC	$kg\ CH_2O$ $kg^{-1}\ DM\ leaf$
ASRQRT	Assimilate requirement for root dry matter production	PLANTC	$kg\ CH_2O$ $kg^{-1}\ DM\ root$
ASRQSA	Assimilate requirement for above-ground storage organ, or crown for sugar beet dry matter production	PLANTC	$kg\ CH_2O$ $kg^{-1}\ DM\ st.\ organ$
ASRQSB	Assimilate requirement for below-ground storage organ dry matter production	PLANTC	$kg\ CH_2O$ $kg^{-1}\ DM\ st.\ organ$
ASRQST	Assimilate requirement for stem dry matter production	PLANTC	$kg\ CH_2O$ $kg^{-1}\ DM\ stem$
ATMTR	Atmospheric transmission coefficient	PENMAN PLANTC RADIAT TOTASS	-

		TOTRAN	
ATRAN	Total actual transpiration rate of the (mixed) canopy	MAIN	mm d^{-1}
		PLANTC	
		TOTRAN	
		WBAL	
AVRAD	Daily incoming total global radiation	MAIN	J m^{-2} ground d^{-1}
		PENMAN	
		PLANTC	
		RADIAT	
		TOTASS	
		TOTRAN	
B	Parameter in Ångström formula	MAIN	-
		PENMAN	
BS	Name for calculation SSLMAX	PLANTC	-
BU	Intermediate variable in calculation of evaporative demand of the atmosphere	PENMAN	s m^{-1}
CNTR	Country name for weather data	MAIN	-
COSLD	Intermediate variable in calculation of daylength	ASTRO	-
		PLANTC	
		RADIAT	
		TOTASS	
CRPF	Crop factor for transpiration	PLANTC	-
		TOTRAN	
DATE	Daynumber since 1 January	PLANTC	d
		WBAL	
DAY	Daynumber since 1 January	ASTRO	d
		MAIN	
		RADIAT	
		TOTASS	
DAYB	Daynumber to start simulation of crop growth	MAIN	d
DAYL	Daylength	ASTRO	h d^{-1}
		PLANTC	
		RADIAT	
		TOTASS	
DAYLP	Photoperiodic daylength	ASTRO	h d^{-1}
		PLANTC	
DEC	Declination of the sun	ASTRO	radians
DELT	Time interval of integration	MAIN	d
		PLANTC	
		WBAL	
DELTA	Derivative of SVPA with respect to temperature	PENMAN	mbar K^{-1}
DLV	Death rate of leaves	PLANTC	kg DM ha^{-1} d^{-1}
DRABS	Total global absorbed radiation	TOTASS	J m^{-2} ground s^{-1}
DRAD	Daily incoming total radiation	MAIN	kJ m^{-2} d^{-1}

DRL	Death rate of leaves	PLANTC	d^{-1}
DSINB	Integral of SINB over the day	RADIAT	$s\ d^{-1}$
DSINBE	As DSINB, but with a correction for lower atmospheric transmission at lower solar elevations	RADIAT	$s\ d^{-1}$
DSOA	Death rate of above-ground storage organs loss	PLANTC	kg DM ha^{-1} ground d^{-1}
DST	Death rate of stems	PLANTC	kg DM ha^{-1} ground d^{-1}
DTGA	Daily total gross CO_2 assimilation of a species	PLANTC	$kg\,CO_2$ ha^{-1} ground d^{-1}
		TOTASS	
DUMMY	Variable to continue the program after a warning	MAIN	-
E0	Penman evaporation of a free water surface	PENMAN	$mm\ d^{-1}$
		TOTRAN	
EA	Evaporative demand of the atmosphere above a bare soil	PENMAN	$mm\ d^{-1}$
EAC	Evaporative demand of the atmosphere above crop stand	PENMAN	$mm\ d^{-1}$
EFF	Initial light use efficiency for individual leaves	ASSIMC	$kg\ CO_2\ ha^{-1}\ leaf\ h^{-1}$ $(J\ m^{-2}\ leaf\ s^{-1})^{-1}$
		PLANTC	
		TOTASS	
EFFTB	Table of EFF versus leaf temperature	PLANTC	$kg\ CO_2\ ha^{-1}\ leaf\ h^{-1}$ $(J\ m^{-2}\ leaf\ s^{-1})^{-1}$
ELV	Height above sea level	MAIN	m
ES0	Penman evaporation of a bare soil	DEVAP	$mm\ d^{-1}$
		MAIN	
		PENMAN	
		PLANTC	
		TOTRAN	
ET0	Penman evaporation of a crop stand	PENMAN	$mm\ d^{-1}$
		PLANTC	
		TOTRAN	
EXDV	Exponent for light intensity calculation (PAR) diffuse	ASSIMC	-
EXSDFV	Exponent for light intensity calculation (PAR) direct component direct flux	ASSIMC	-
EXSTV	Exponent for light intensity calculation (PAR) total direct flux	ASSIMC	-
FAG	Fraction of total dry matter increase allocated to above-ground plant parts (= shoot)	PLANTC	-
FAGTB	Table of FAG versus temperature sum (or DVS)	PLANTC	-
FAI	Flower area index	ASSIMC	m^2 flower
		PLANTC	m^{-2} ground
		TOTASS	

FAIC	Flower area index above selected height in canopy	ASSIMC	m^2 flower m^{-2} ground
		PLANTC	
FBG	Fraction of total dry matter increase allocated to below-ground plant parts	PLANTC	-
FD	Flower area density at selected height	ASSIMC	m^2 flower m^{-3}
FGL	Assimilation rate at selected canopy height	ASSIMC	kg CO_2 ha^{-1}ground h^{-1}
FGROS	Canopy assimilation summed over layers	ASSIMC	kg CO_2
		TOTASS	ha^{-1} ground h^{-1}
FGRS	Assimilation of sunlit leaves per leaf angle	ASSIMC	kg CO_2 ha^{-1} leaf h^{-1}
FGRSH	Assimilation of shaded leaves	ASSIMC	kg CO_2 ha^{-1} leaf h^{-1}
FGRSHF	Assimilation of shaded flowers	ASSIMC	kg CO_2 ha^{-1} leaf h^{-1}
FGRSHS	Assimilation of shaded stems	ASSIMC	kg CO_2 ha^{-1} leaf h^{-1}
FGRSUF	Instantaneous CO_2 assimilation rate of sunlit flowers	ASSIMC	kg CO_2 ha^{-1} leaf h^{-1}
FGRSUN	Instantaneous CO_2 assimilation rate of sunlit leaves averaged over leaf angles	ASSIMC	kg CO_2 ha^{-1}leaf h^{-1}
FGRSUS	Instantaneous CO_2 assimilation rate of sunlit stems	ASSIMC	kg CO_2 ha^{-1} leaf h^{-1}
FILEO	File name for output variables	MAIN	-
FILER	File name for rerun variables	MAIN	-
FILES	File name for soil variables	MAIN	-
		WBAL	
FILET	File name for time variables	MAIN	-
		PLANTC	
FINTIM	Period of simulation	MAIN	d
FLV	Fraction of FAG allocated to leaves	PLANTC	-
FLVTB	Table of FLV versus temperature sum (or DVS)	PLANTC	-, °C d
FRABS	Fraction absorbed of incoming global radiation	PLANTC	-
		TOTASS	
		TOTRAN	
FRD	Fraction of global radiation reaching the ground used for calculation of soil evaporation	DEVAP	-
		MAIN	
		PLANTC	
		TOTASS	
FRDIF	Diffuse light fraction from atmospheric transmission	RADIAT	-
FRT	Fraction of FBG allocated to roots	PLANTC	-
FRTTB	Table of FRT versus temperature sum (or DVS)	PLANTC	-, °C d
FSLLAV	Fraction sunlit leaf area for a species in leaf layer X (PAR)	ASSIMC	-
FSOA	Fraction of FAG allocated to above-ground storage organs, or crowns for sugar beet	PLANTC	-
FSOATB	Table of FSOA versus temperature sum (or DVS)	PLANTC	-, °C d
FSOB	Fraction of FBG allocated to below-ground storage organs	PLANTC	-

FST	Fraction of FAG allocated to stem, or petioles for sugar beet	PLANTC	-
FSTTB	Table of FST versus temperature sum (or DVS)	PLANTC	-, °C d
GAG	Growth rate of above-ground plant parts	PLANTC	kg DM ha^{-1} ground d^{-1}
GAMMA	Psychrometric constant	PENMAN	mbar K^{-1}
GLAI	Growth rate of leaf area index	PLANTC	ha^{-1} leaf ha^{-1} ground d^{-1}
GLV	Gross growth rate of leaves	PLANTC	kg DM ha^{-1} ground d^{-1}
GPHOT	Daily total gross CH_2O assimilation of a species	PLANTC	kg CH_2O ha^{-1} ground d^{-1}
GRLV	Net growth rate of leaves	PLANTC	kg DM ha^{-1} ground d^{-1}
GRSOA	Growth rate of above-ground storage organs, or crowns for sugar beet	PLANTC	kg DM ha^{-1} ground d^{-1}
GRST	Growth rate of stems, or petioles for sugar beet	PLANTC	kg DM ha^{-1} ground d^{-1}
GRT	Growth rate of roots	PLANTC	kg DM ha^{-1} ground d^{-1}
GSOB	Growth rate of below-ground storage organs	PLANTC	kg DM ha^{-1} ground d^{-1}
GTW	Total growth rate of a species	PLANTC	kg DM ha^{-1} ground d^{-1}
HARDAY	Julian daynumber at which output is required to compare	MAIN	-
HB	Height growth parameter	PLANTC	-
HGHT	Total height of the photosynthetic surface of a species in the canopy	ASSIMC LEAFPA LEAFRE PLANTC TOTASS	m
HGHTI	Initial height of the photosynthetic surface of a species in the canopy	PLANTC	m
HMAX	Height growth parameter (potential plant height)	PLANTC	m
HOUR	Selected hour during the day at which instantaneous CO_2 assimilation rate of each species is calculated	RADIAT TOTASS	h
HS	Logistic height growth parameter	PLANTC	°C^{-1} d^{-1}
HU	Daily heat unit for plant development	PLANTC	°C
HULV	Daily heat unit for leaf development	PLANTC	°C
I	Counter	ASSIMC TOTASS	-
I1	Counter for reruns	MAIN	-
I2	Length string	MAIN	-

I3	Length string	MAIN	-
IDAY	Integer variable for day number since 1 January	MAIN PLANTC	d
IDAYEM	Integer variable for day of crop emergence	PLANTC	d
IDTMP	Switch variable for temporary output file	MAIN	-
IG1	Counter	ASSIMC	-
IG2	Counter	ASSIMC	-
IGP1	Number of points in Gaussian integration	ASSIMC	-
IGP2	Number of points in Gaussian integration	ASSIMC	-
ILAMD	Number of elements of the array AMDVST	PLANTC	-
ILEFF	Number of elements of the array EFFTB	PLANTC	-
ILFAG	Number of elements of the array FAGTB	PLANTC	-
ILFLV	Number of elements of the array FLVTB	PLANTC	-
ILFRT	Number of elements of the array FRTTB	PLANTC	-
ILFSOA	Number of elements of the array FSOATB	PLANTC	-
ILFST	Number of elements of the array FSTTB	PLANTC	-
ILLAI	Number of elements of the array LAITB	PLANTC	-
ILPTB	Number of elements of the array PTB	PLANTC TOTRAN	-
ILRDRO	Number of elements of the array RDRSOT	PLANTC	-
ILRDRS	Number of elements of the array RDRSTT	PLANTC	-
ILREDF	Number of elements of the array REDFTB	PLANTC	-
ILREDM	Number of elements of the array REDMNT	PLANTC	-
ILSLA	Number of elements of the array SLATB	PLANTC	-
IMAX	Maximum number of species	ASSIMC TOTASS	-
IMNHD	Maximum number of harvest dates	MAIN	-
IMNP	Integer for declaration maximum length of array	PLANTC TOTRAN	-
IMNS	Integer for declaration maximum number of species	PLANTC TOTRAN	-
INFR	Infiltration of water into the soil, as a sum of rain and irrigation	WBAL	mm d^{-1}
INGP	Number (n) of times during a day that the instantaneous CO_2 assimilation rate for each species is calculated (for Gaussian integration)	TOTASS	-
INHD	Array length of harvest dates	MAIN	-
INIT	Variable which indicates whether soil variables should be used	WBAL	-
INITC	Control variable for putting variable values in output file	PLANTC	-
INITS	Control variable for putting variable values in output file	PLANTC	-
INS	Number of species	ASSIMC	-

		PLANTC	
		TOTASS	
		TOTRAN	
INS1	Help variable (number of species)	PLANTC	-
INS2	Help variable (number of species)	PLANTC	-
INSETS	Number of rerun sets	MAIN	-
IPSPEC	Species number	PLANTC	-
IRCAP	Number of values found on array RCAPT	WBAL	-
IRIRR	Number of values found on array RIRRIT	WBAL	-
IRUN	Number of running program	MAIN	-
IRUNLA	Integer for making choice using measured LAI	PLANTC	-
	as input (1) or using LAI simulated (0)		
IS	Counter (species number)	PLANTC	-
		TOTRAN	
ISTAT1	Help variable	MAIN	-
ISTAT2	Help variable	MAIN	-
ISTN	Integer number to refer to selected	MAIN	-
	meteorological station		
ITABLE	Format of output file	MAIN	-
ITASK	Integer determining what subroutine should do	MAIN	-
	(1 = initialization, 2 = rate calculation,	PLANTC	
	3 = integration, 4 = terminal)	WBAL	
ITOLD	Integer determined by the task that was	PLANTC	-
	done during the previous call to PLANTC	WBAL	
IUNITO	Unit number for output file	MAIN	-
		PLANTC	
		WBAL	
IUNITP	Unit number for plant data file	MAIN	-
		PLANTC	
IUNITR	Unit number for rerun file	MAIN	-
IUNITS	Unit number for soil data file	MAIN	-
		WBAL	
IUNITT	Unit number for timer file	MAIN	-
IYEAR	Integer for the year for which weather data	MAIN	-
	are requested		
K	Counter	ASSIMC	-
		TOTASS	
KDF	Extinction coefficient for leaves	ASSIMC	-
		PLANTC	
		TOTASS	
KDFFV	Extinction coefficient for PAR diffuse flux flowers	ASSIMC	-
KDFSV	Extinction coefficient for PAR diffuse flux stems	ASSIMC	-
KDFV	Extinction coefficient for PAR diffuse flux leaves	ASSIMC	-
KDRBFV	Extinction coefficient for PAR direct component	ASSIMC	-

	direct flux flowers		
KDRBLV	Extinction coefficient for PAR direct component	ASSIMC	-
	direct flux leaves		
KDRBSV	Extinction coefficient for PAR direct component	ASSIMC	-
	direct flux stems		
KDRTFV	Extinction coefficient for PAR total direct flux flowers	ASSIMC	-
KDRTSV	Extinction coefficient for PAR total direct flux stems	ASSIMC	-
KDRTV	Extinction coefficient for PAR total direct flux leaves	ASSIMC	-
KF	Extinction coefficient for flowers	ASSIMC	-
		PLANTC	
		TOTASS	
KS	Extinction coefficient for stems	ASSIMC	-
		PLANTC	
		TOTASS	
LA0	Initial leaf area (at field emergence)	PLANTC	$cm^2 plant^{-1}$
LAI	Leaf area index	ASSIMC	ha leaf ha^{-1} ground
		LEAFPA	
		LEAFRE	
		PLANTC	
		TOTASS	
LAIC	Leaf area index above selected height in canopy	ASSIMC	ha leaf ha^{-1} ground
		LEAFRE	
		LEAFPA	
LAID	Leaf area index at the present day	PLANTC	ha leaf ha^{-1} ground
LAIM	Maximum value of leaf area index	PLANTC	ha leaf ha^{-1} ground
LAITB	Table of LAI versus daynumber	PLANTC	ha leaf ha^{-1} ground, d
LAITOT	Total leaf area index of species	PLANTC	ha leaf ha^{-1} ground
		TOTASS	
LAIY	Leaf area index at the day before	PLANTC	ha leaf ha^{-1} ground
LAT	Latitude of the weather station	ASTRO	degrees
		MAIN	
		PLANTC	
LD	Leaf area density at point X in the canopy	ASSIMC	$m^2 leaf m^{-3}$
		LEAFPA	
		LEAFRE	
LHVAP	Latent heat of evaporation of water	PENMAN	$J kg^{-1} H_2O$
LONG	Longitude of the weather station	MAIN	degrees
MAINLV	Maintenance respiration coefficient of leaves	PLANTC	$kg CH_2O kg^{-1} DM d^{-1}$
MAINRT	Maintenance respiration coefficient of roots	PLANTC	$kg CH_2O kg^{-1} DM d^{-1}$
MAINSA	Maintenance respiration coefficient of above-ground storage organs	PLANTC	$kg CH_2O kg^{-1} DM d^{-1}$

MAINSB	Maintenance respiration coefficient of below-ground storage organs	PLANTC	kg CH_2O kg^{-1} DM d^{-1}
MAINST	Maintenance respiration coefficient of stems	PLANTC	kg CH_2O kg^{-1} DM d^{-1}
MAINT	Maintenance respiration rate of a species	PLANTC	kg CH_2O ha^{-1} d^{-1}
MAINTS	Maintenance respiration rate of a species at reference temperature	PLANTC	kg CH_2O ha^{-1} d^{-1}
MNDVS	Coefficient to account for lower respiration when plants are ageing	PLANTC	-
NPL	Plant density	PLANTC	plants m^{-2}
OUTPUT	Logical for call OUTDAT routine	MAIN PLANTC WBAL	-
P	Soil depletion factor	TOTRAN	-
PAR	Instantaneous flux of photosynthetic active radiation	RADIAT	J m^{-2} ground s^{-1}
PARDIF	Instantaneous flux of diffuse PAR	ASSIMC RADIAT TOTASS	J m^{-2} ground s^{-1}
PARDIR	Instantaneous flux of direct PAR	ASSIMC RADIAT TOTASS	J m^{-2} ground s^{-1}
PBAR	Barometric pressure	PENMAN	mbar
PERC	Percolation of water from top layer to root zone	WBAL	mm d^{-1}
PI	Ratio of circumference to diameter of circle	ASTRO RADIAT TOTASS	-
PRDEL	Time interval for printing	MAIN	d
PSYCON	Psychrometric instrument constant	PENMAN	degree K^{-1}
PTB	Table of P versus ET0	PLANTC TOTRAN	-, mm d^{-1}
PTRAN	Potential transpiration rate derived from Penman evaporation	TOTRAN	mm d^{-1}
Q10	Factor accounting for increase of maintenance respiration with a 10 $^{\circ}$C rise temperature	PLANTC	-
RAD	Factor to convert degrees into radians	ASTRO	radians $degree^{-1}$
RADABS	Absorbed radiation (PAR) by species in canopy at selected time of the day	ASSIMC TOTASS	J m^{-2} ground d^{-1}
RADDIF	Incoming global diffuse radiation	ASSIMC TOTASS	J m^{-2} ground s^{-1}
RADDIR	Incoming global direct radiation	ASSIMC TOTASS	J m^{-2} ground s^{-1}
RAIN	Water input through rainfall	MAIN WBAL	mm d^{-1}
RAINT	Total water input through rainfall over simulation	WBAL	mm

	period		
RB	Net outgoing long wave radiation according to Brunt (1932)	PENMAN	J m^{-2} ground d^{-1}
RCAP	Water input through capillary rise of soil water	WBAL	mm d^{-1}
RCAPT	Table of RCAP versus daynumber	WBAL	mm d^{-1}, d
RDRAIN	Percolation of water from root layer to deeper parts of the soil	WBAL	mm d^{-1}
RDRSOA	Relative rate of loss of seeds	PLANTC	d^{-1}
RDRSOT	Table of RDRSOA versus temperature sum (or DVS)	PLANTC	d^{-1}, °C d
RDRST	Relative death rate of stems, or petioles for sugar beet	PLANTC	(°C d)$^{-1}$
RDRSTT	Table of RDRST versus temperature sum (or DVS)	PLANTC	(°C d)$^{-1}$, °C d
RDUM	Dummy variable	PLANTC	-
REDF	Factor accounting for effect of temperature on AMAX	PLANTC	-
REDFS	Factor accounting for effect of relative water content of the top layer of the soil on AEVAP	DEVAP	-
REDFST	Table of REDFS versus WCPR (relative water content)	DEVAP	-, -
REDFTB	Table of REDF versus temperature	PLANTC	-, °C
REDIST	Redistribution of dry matter from yellowing above-ground plant parts	PLANTC	kg DM ha^{-1} ground d^{-1}
REDLM	Redistribution coefficient for yellowing leaves	PLANTC	-
REDMN	Factor accounting for effect of minimum temperature on AMAX	PLANTC	-
REDMNT	Table of REDMN versus minimum temperature	PLANTC	-, °C
REDST	Redistribution coefficient for yellowing stems	PLANTC	-
REFCFC	Reflection coefficient for a canopy surface	PENMAN	-
REFCFS	Reflection coefficient for a soil surface	PENMAN	-
REFCFW	Reflection coefficient for a water surface	PENMAN	-
REFVH	Reflection coefficient of canopy with horizontal leaves (for PAR)	ASSIMC	-
REFVS	Reflection coefficient of canopy with spherical leaf angel distribution (for PAR)	ASSIMC	-
REFTMP	Reference temperature for maintenance respiration	PLANTC	°C
RELSSD	Hours of sunshine per daylength (variable in the Penman formula)	PENMAN	h d^{-1}
RGRL	Relative growth rate of leaf area	PLANTC	(°C d)$^{-1}$
RHG	Rate of height growth	PLANTC	cm d^{-1}
RIRRI	Water input through irrigation	WBAL	mm d^{-1}
RIRRIT	Table of RIRRI versus daynumber	WBAL	mm d^{-1}, d
RNC	Net absorbed radiation of a crop in Penman calculation	PENMAN	J m^{-2} ground d^{-1}
RNS	Net absorbed radiation of a bare soil in Penman calculation	PENMAN	J m^{-2} ground d^{-1}

RNW	Net absorbed radiation of a free water surface in Penman calculation	PENMAN	$J\ m^{-2}$ water d^{-1}
RTD	Rooted depth	MAIN PLANTC WBAL	m
RTDMAX	Maximum rooted depth	WBAL	m
RTDT	Thickness rooted layer	PLANTC WBAL	m
SAI	Stem area index	ASSIMC PLANTC TOTASS	m^2 stem m^{-2} ground
SAIC	Stem area index above selected height in canopy	ASSIMC	m^2 stem m^{-2} ground
SC	Solar constant, corrected for varying distances between sun and earth	RADIAT	$J\ m^{-2}\ s^{-1}$
SCV	Scattering coefficients of leaves (PAR)	ASSIMC	-
SD	Stem area density at point X in the canopy	ASSIMC	m^2 stem m^{-3} canopy
SFA	Specific flower area	PLANTC	m^2 flower kg^{-1}
SINB	Sine of solar elevation	ASSIMC RADIAT TOTASS	-
SINLD	Intermediate variable in calculating daylength	ASTRO PLANTC RADIAT TOTASS	-
SLA	Specific leaf area	PLANTC	ha leaf kg^{-1} leaf
SLAC	Calculated SLA early growth	PLANTC	ha leaf kg^{-1} leaf
SLAMAX	Maximum SLA	PLANTC	ha leaf kg^{-1} leaf
SLATB	Table of SLA versus temperature sum	PLANTC	ha leaf kg^{-1} leaf, $^{\circ}$C d
SMFCT	Soil moisture content at field capacity of the top layer of the soil	WBAL	mm
SMFRTZ	Soil moisture content at field capacity of root layer	WBAL	mm
SMRTD	Actual soil moisture content of root and top layer	WBAL	mm
SMRTZ	Actual soil moisture content of root layer	WBAL	mm
SMRTZI	Initial soil moisture content of root layer at the start of simulation	WBAL	mm
SMT	Actual soil moisture content of the top layer of the soil	WBAL	mm
SMTI	Initial soil moisture content of the top layer of the soil at the start of simulation	WBAL	mm
SSA	Specific stem area	PLANTC	$m^2\ kg^{-1}$
SSL	Specific stem length	PLANTC	cm kg^{-1}
SSLMAX	Maximum value of SSL	PLANTC	cm kg^{-1}

STBC	Stefan-Boltzmann constant	PENMAN	J m^{-2} d^{-1} K^{-1}
SVPA	Saturated vapour pressure in the air	PENMAN	mbar
TAEVAP	Total soil evaporation integrated in time	WBAL	mm
TATRAN	Total crop transpiration integrated in time	WBAL	mm
TCS	Time coefficient for delay in drainage	WBAL	d
TDIF	Maximum daily temperature difference	PENMAN	°C
TDRABS	Total global absorbed radiation integrated in time	TOTASS	J m^{-2} ground s^{-1}
TEFF	Factor accounting for effect of temperature on maintenance respiration	PLANTC	-
TERMNL	Variable which indicates whether the simulation loop should be terminated	PLANTC MAIN WBAL	-
THCKT	Thickness of the top layer	WBAL	m
TIME	Daynumber start simulation	MAIN PLANTC	d
TINY	Help parameter for generating output on harvest day	MAIN	-
TMD	Base temperature for plant development	PLANTC	°C
TMDLV	Base temperature for leaf development	PLANTC	°C
TMN	Daily minimum air temperature	MAIN PENMAN PLANTC TOTRAN	°C
TMPA	Daily average air temperature	PENMAN PLANTC	°C
TMTMX	Average temperature in daytime	PLANTC	°C
TMX	Daily maximum air temperature	MAIN PENMAN PLANTC TOTRAN	°C
TMXD	Maximum temperature for phenological development	PLANTC	°C
TMXLV	Maximum temperature for leaf area development	PLANTC	°C
TRAN	Actual transpiration rate of a species	PLANTC TOTRAN	mm d^{-1}
TRANRF	Factor accounting for effect of water stress on the rate of dry matter increase	PLANTC TOTRAN	-
TS	Temperature sum for plant development	PLANTC	°C d
TSLAM	Temperature sum at which leaf death starts	PLANTC	°C d
TSLV	Temperature sum for leaf development	PLANTC	°C d
VAPOUR	Average vapour pressure in the air	MAIN PENMAN PLANTC TOTRAN	mbar (from kPa)
VISSUN	Absorbed flux for each species for sunlit leaves	ASSIMC	J m^{-2} leaf s^{-1}
VSMAD	Volumetric soil moisture content at air dryness	DEVAP	dm^3 water m^{-3} soil

		MAIN	
		WBAL	
VSMCR	Critical volumetric soil moisture content below which	TOTRAN	dm^3 water m^{-3} soil
	the rate of dry matter increase is reduced as a result		
	of water stress		
VSMFC	Volumetric soil moisture content at field capacity	DEVAP	dm^3 water m^{-3} soil
	soil	MAIN	
		PLANTC	
		TOTRAN	
		WBAL	
VSMRTZ	Actual volumetric soil moisture content of root layer	MAIN	dm^3 water m^{-3} soil
		PLANTC	
		TOTRAN	
		WBAL	
VSMT	Actual volumetric soil moisture content of the	DEVAP	dm^3 water m^{-3} soil
	top layer of the soil	MAIN	
		WBAL	
VSMWP	Volumetric soil moisture content at wilting point	MAIN	dm^3 water m^{-3} soil
		PLANTC	
		TOTRAN	
		WBAL	
WAG	Dry weight of total green shoot (field experiments)	PLANTC	kg DM ha^{-1}
WCPR	Relative water content of the top layer of the soil	DEVAP	dm^3 water m^{-3} soil
WDTOT	Total crop dry weight	PLANTC	kg DM ha^{-1}
WGAUSS	Weights of points in Gaussian integration	TOTASS	-
WGAUS1	Weights of points in Gaussian integration	ASSIMC	-
WGAUS2	Weights of points in Gaussian integration	ASSIMC	-
WIND	Daily average wind speed	MAIN	$m\ s^{-1}$
		PENMAN	
		PLANTC	
		TOTRAN	
WLVD	Dry weight of dead leaves (experiments)	PLANTC	kg DM ha^{-1}
WLVG	Dry weight of green leaves	PLANTC	kg DM ha^{-1}
WLVGI	Initial dry weight of green leaves	PLANTC	g plant^{-1}
WLVGM	Maximum dry weight of green leaves	PLANTC	kg DM ha^{-1}
WRT	Dry weight of roots	PLANTC	kg DM ha^{-1}
WRTI	Initial dry weight of roots	PLANTC	g plant^{-1}
WSOA	Dry weight of above-ground storage organs, or	PLANTC	kg DM ha^{-1}
	crowns for sugar beet		
WSOB	Dry weight of below-ground storage organs	PLANTC	kg DM ha^{-1}
WSOD	Dry weight of dead storage organs	PLANTC	kg DM ha^{-1}
WSTD	Dry weight of dead stems, or petioles for sugar beet	PLANTC	kg DM ha^{-1}
WSTG	Dry weight of green stems, or petioles for sugar beet	PLANTC	kg DM ha^{-1}
WSTGI	Initial dry weight of green stems, or petioles for	PLANTC	g plant^{-1}

	sugar beet		
WTOT	Total green crop dry weight, including below-ground plant parts	PLANTC	kg DM ha^{-1}
WTRDIR	Directory and path weather files	MAIN	-
WTRMES	Flag for messages from the weather system	MAIN	-
WTROK	Help variable	MAIN	-
X	Selected height	ASSIMC	m
		LEAFPA	
		LEAFRE	
XGAUSS	Points for Gaussian integration	TOTASS	-
XGAUS1	Points for Gaussian integration	ASSIMC	-
XGAUS2	Points for Gaussian integration	ASSIMC	-
YLV	Yellowing rate of leaves	PLANTC	kg DM ha^{-1} d^{-1}
YST	Yellowing rate of stems	PLANTC	kg DM ha^{-1} d^{-1}

Appendix Four

List of Symbols Used in Equations

Symbol	Description	Eqn first used	Code in INTERCOM model
a	Constant in function for calculation of SSL_{max} (m^{-1})	4.41	AS
a_0	Coefficient in yield - density relationship	2.4	-
a_A	Constant in Ångström formula (-)	5.3	A
A_d	Daily CO_2 assimilation rate of a canopy (kg CO_2 ha^{-1} ground d^{-1})	4.26	DTGA
A_h	Actual rate of CO_2 assimilation at height h in the canopy (kg CO_2 ha^{-1} leaf h^{-1})	4.20	FGRSH, FGRSUN
A_m	Maximum CO_2 assimilation rate (at light saturation) (kg CO_2 ha^{-1} leaf h^{-1})	4.20	AMAX
b	Height growth parameter (-)	4.40	HB
b_A	Constant in Ångström formula (-)	5.3	B
b_0	Coefficients used in hyperbolic yield - density relationships (-)	2.1	-
b_c	-	2.1	-
b_{c0}	-	2.2	-
b_{cc}	-	2.2	-
b_{cw}	-	2.2	-
b_{w0}	-	2.2	-
b_{wc}	-	2.2	-
b_{ww}	-	2.2	-
c	Constant in function for calculation of SSL_{max} (-)	4.41	BS
c_i	Proportionality factor for evapotranspiration (relation to short grass cover) (-)	5.13	CRPF
CR	Capillary rise (mm d^{-1})	5.2	RCAP
D	Development stage (-)	-	DVS

D_N	Demand of the crop for nitrogen (kg N ha^{-1})	6.2	-
D_r	Development rate (d^{-1})	4.30	DVR
$DLAI$	Leaf area death rate (m^2 m^{-2} ($^\circ$C d)$^{-1}$) or (m^2 m^{-2} d^{-1})	4.37	
E	Soil evaporation (kg H$_2$O m^{-2} d^{-1}) or (mm d^{-1})	5.1	AEVAP
E_a	Actual soil evaporation (kg H$_2$O m^{-2} d^{-1}) or (mm d^{-1})	5.1	AEVAP
e_a	Actual vapour pressure (mbar)	5.4	VAPOUR
E_d	Radiation driven component of evapotranspiration (kg H$_2$O m^{-2} d^{-1})	5.9	-
E_{dm}	Average light use efficiency (kg dry matter MJ^{-1})	4.28	-
E_s	Aerodynamic component of evapotranspiration (kg H$_2$O m^{-2} d^{-1})	5.9	-
e_s	Saturated vapour pressure (mbar)	5.8	SVPA
E_p	Potential soil evaporation (kg H$_2$O m^{-2} d^{-1}) or (mm d^{-1})	5.12	-
$E_{p,r}$	Potential soil evaporation of a bare soil (kg H$_2$O m^{-2} d^{-1}) or (mm d^{-1})	5.12	ES0
ET_r	Reference evapotranspiration of a short grass cover (kg H$_2$O m^{-2} d^{-1}) or (mm d^{-1})	5.13	ET0
F	Rate of nitrogen fertilizer (kg N ha^{-1} d^{-1})	6.1	-
FAD	Flower area density (m^2 flower m^{-2} ground m^{-1} height)	4.22	FD
FAI	Flower area index (m^2 reproductive organs m^{-2} ground)	4.22	FAI
f_a	Fraction of radiation absorbed by a species (-)	5.13	FRABS
F_h	Cumulative flower area index above height h (m^2 flower m^{-2} ground)	4.23	FAIC
f_s	Fraction in storage organs which requires no maintenance	4.24	-
f_{sl}	Fraction of sunlit leaf area (-)	4.21	FSLLA
G_a	Actual growth rate of the crop (kg dry matter ha^{-1} d^{-1})	5.16	GTW
G_{lv}	Actual growth rate of the leaves (kg dry matter ha^{-1} d^{-1})	4.36	GRLV
G_p	Potential growth rate of the crop (kg dry matter ha^{-1} d^{-1})	4.26	-
$GLAI$	Leaf area growth rate (m^2 leaf m^{-2} ground d^{-1})	4.36	GLAI

h_0	Lower boundary of leaf (or flower) area (m)	4.22	-
H	Sensible heat flux (J m^{-2} d^{-1})	5.7	-
h	Specific height in the canopy (m)	4.8	-
h_m	Maximum plant height (m)	4.39	HMAX
h_t	Total height of a species (m)	4.7	HGHT
h_u	Wind function (kg m^{-2} d^{-1} mbar^{-1})	5.9	
I_0	Visible incoming radiation flux (*PAR*) at the top of the canopy (J m^{-2} ground s^{-1})	4.5	PAR
$I_{0,df}$	Diffuse component of incoming *PAR* (J m^{-2} ground s^{-1})	4.15	PARDIF
$I_{0,dr}$	Direct component of incoming *PAR* (J m^{-2} ground s^{-1})	4.16	PARDIR
I_a	Absorbed radiation flux by a leaf (J m^{-2} leaf s^{-1})	4.20	-
$I_{a,df}$	Absorbed flux of diffuse radiation (J m^{-2} leaf s^{-1})	4.15	AFVV
$I_{a,dr,dr}$	Absorbed flux of the direct component of direct radiation (J m^{-2} leaf s^{-1})	4.17	AFVD
$I_{a,dr,t}$	Absorbed flux of the total direct radiation (J m^{-2} leaf s^{-1})	4.16	AFVT
$I_{a,sh}$	Absorbed flux of shaded leaves (J m^{-2} leaf s^{-1})	4.18	AFVSHD
I_h	Incoming visible radiation flux at height h in the canopy (J m^{-2} ground s^{-1})	4.6	-
I_L	Downward flux of radiation (*PAR*) at depth L in the canopy (with an *LAI* of L above that point) (J m^{-2} ground s^{-1})	4.5	-
IR	Irrigation rate (mm)	5.1	RIRRI
k	Extinction coefficient for visible radiation (-)	4.5	-
k_{df}	Extinction coefficient for diffuse visible radiation (-)	4.13	KDFV
$k_{dr,bl}$	Extinction coefficient for direct component of direct visible radiation (-)	4.13	KDRBLV
$k_{dr,t}$	Extinction coefficient for total direct visible radiation (-)	4.14	KDRTV
l	Effective root length (m ha^{-1})	6.5	-
L	*LAI* above depth L in the canopy (m^2 leaf m^{-2} ground)	4.5	LAIC
$L_{p,0}$	Initial leaf area per plant at seedling emergence (m^2 leaf plant^{-1})	4.33	LA0
LAD_h	Leaf area density at height h (m^2 leaf m^{-2} ground m^{-1} height)	4.7	LD

L_h	Cumulative leaf area index counted from the maximum height of the species to height h (m^2 leaf m^{-2} ground)	4.6	LAIC
L_p	Leaf area per plant (m^2 plant^{-1})	4.34	-
L_w	Relative leaf area index of the weeds	2.16	-
LA_c	Leaf area per crop plant (m^2 plant^{-1})	2.15	-
LA_w	Leaf area per weed plant (m^2 plant^{-1})	2.15	-
LAI	Leaf area index (m^2 leaf m^{-2} ground)	4.7	LAI
LAI_c	Leaf area index of the crop	2.13	-
LAI_w	Leaf area index of the weeds	2.13	-
m	Maximum yield loss in yield loss - weed density equation (-)	2.8	-
M	Rate of net mineralization of N (kg N ha^{-1} d^{-1})	6.1	-
mc_{lv}	Coefficient for maintenance respiration of leaves (kg CH$_2$O kg^{-1} dry matter d^{-1})	4.24	MAINLV
mc_{rt}	Coefficient for maintenance respiration of roots (kg CH$_2$O kg^{-1} dry matter d^{-1})	4.24	MAINRT
mc_{so}	Coefficient for maintenance respiration of storage organs (kg CH$_2$O kg^{-1} dry matter d^{-1})	4.24	MAINSO
mc_{st}	Coefficient for maintenance respiration of stems (kg CH$_2$O kg^{-1} dry matter d^{-1})	4.24	MAINST
N	Number of plants (plants m^{-2})	4.33	NPL
N_c	Plant density of the crop (plants m^{-2})	2.1	-
N_c	Nitrogen content of the crop (kg N ha^{-1})	6.2	-
n_s	Actual number of sunshine hours	5.3	-
N_s	Maximum number of sunshine hours	5.3	-
N_s	Mineral nitrogen content of the soil (kg N ha^{-1})	6.1	-
N_w	Plant density of the weeds (plants m^{-2})	2.2	-
NC_a	Actual nitrogen concentration (kg N kg^{-1} dry matter)	6.4	-
NC_{cr}	Critical nitrogen concentration (kg N kg^{-1} dry matter)	6.4	-
NC_m	N concentration of the crop (kg N kg^{-1} dry matter)	6.2	
NC_{mn}	Minimum nitrogen concentration (kg N kg^{-1} dry matter)	6.4	-
NIR	Near infra-red radiation (J m^{-2} ground s^{-1})	-	-
p	Soil moisture depletion factor (-)	5.15	P
P	Percolation (kg H$_2$O m^{-2} d^{-1})	5.1	PERC,
		5.2	RDRAIN

PAR	Photosynthetically active radiation (wavelength 400 - 700 nm) ($J\ m^{-2}$ ground s^{-1})	-	-
pc_k	Partitioning coefficient (kg dry matter (plant organ k) / kg dry matter crop)	7.2	FLV,FRT FST, FSOA, FSOB
q	Relative damage coefficient (-)	2.13	-
q_0	Relative damage coefficient (-)	2.21	-
Q	Assimilate requirement for dry matter production (kg $CH_2O\ kg^{-1}$ dry matter)	4.26	ASRQ
R	Rainfall (kg $H_2O\ m^{-2}\ d^{-1}$) or (mm d^{-1})	5.1	RAIN
r	Recovery of nitrogen fertilizer (-)	6.1	-
R_b	Net outgoing long-wave radiation ($J\ m^{-2}$ ground d^{-1})	5.4	RB
R_l	Relative growth rate of leaf area ($^{\circ}C^{-1}\ d^{-1}$)	4.33	RGRL
R_m	Maintenance respiration at actual temperature (kg $CH_2O\ ha^{-1}\ d^{-1}$)	4.25	MAINT
$R_{m,r}$	Maintenance respiration at reference temperature (kg $CH_2O\ ha^{-1}\ d^{-1}$)	4.24	MAINTS
R_n	Net radiation ($J\ m^{-2}$ ground d^{-1})	5.7	RNC
R_s	Relative death rate of leaves ($^{\circ}C^{-1}\ d^{-1}$)	4.37	DRL
s	Slope of the saturated vapour pressure curve (mbar $^{\circ}C^{-1}$)	5.9	DELTA
s	Logistic height growth parameter ($^{\circ}C^{-1}\ d^{-1}$)	4.40	HS
S_0	Total theoretical global radiation without atmosphere ($J\ m^{-2}$ ground s^{-1})	4.1	-
$S_{0,d}$	Daily total theoretical global radiation without atmosphere ($J\ m^{-2}$ ground d^{-1})	4.2	ANGOT
S_g	Total global radiation, measured ($J\ m^{-2}$ ground s^{-1})	4.3	-
$S_{g,d}$	Daily total global radiation, measured ($J\ m^{-2}$ ground d^{-1})	4.3	AVRAD
$S_{g,df,d}$	Daily diffuse radiation ($J\ m^{-2}$ ground d^{-1})	Fig. 4.1	PARDIF
S_{sc}	Solar constant, incoming global radiation without an atmosphere, perpendicular to sun rays ($J\ m^{-2}$ ground s^{-1})	4.1	SC
SAI	Stem area index (m^2 stem m^{-2} ground)	-	SAI
sh_i	Effective share in leaf area of species i (-)	4.27	-
SLA	Specific leaf area (m^2 leaf kg^{-1} leaf)	4.36	SLA

SSL	Specific stem length (m stem kg^{-1} dry matter)	-	SSL
SSL_{max}	Maximum specific stem length (m stem kg^{-1} dry matter)	4.41	SSLMAX
T_a	Actual transpiration by the vegetation ($kg\ H_2O\ m^{-2}\ d^{-1}$) or ($mm\ d^{-1}$)	5.2	ATRAN
T_{av}	Average daily temperature (°C)	4.25	TMPA
T_b	Base temperature for development (°C)	4.29	TMD
T_c	Time coefficient for nitrogen uptake (d)	6.2	-
T_{cw}	Period between crop and weed emergence (d)	2.9	-
T_p	Potential transpiration of a canopy ($kg\ H_2O\ m^{-2}\ d^{-1}$) or ($mm\ d^{-1}$)	5.14	PTRAN
T_r	Reference temperature for maintenance respiration (°C)	4.25	REFTMP
ts	Temperature sum (°C d)	4.29	TS
U	Uptake rate of mineral nitrogen ($kg\ N\ ha^{-1}\ d^{-1}$)	6.1	-
u_2	Wind speed at 2 m height ($m\ s^{-1}$)	5.10	WIND
W	Weight of the crop ($kg\ ha^{-1}$)	6.2	-
W_c	Weight of a crop plant ($g\ plant^{-1}$)	2.1	-
W_w	Weight of a weed plant ($g\ plant^{-1}$)	2.2	-
W_{lv}	Weight of the leaves ($kg\ ha^{-1}$)	4.24	WLVG
W_{rt}	Weight of the roots ($kg\ ha^{-1}$)	4.24	WRT
W_{so}	Weight of the storage organs ($kg\ ha^{-1}$)	4.24	WSOA WSOB
W_{st}	Weight of the stems ($kg\ ha^{-1}$)	4.24	WSTG
x	Coefficient in yield loss equation	2.9	-
y	Coefficient in yield loss equation	2.9	-
Y_{cm}	Yield of the crop in monoculture ($g\ m^{-2}$)	2.1	-
Y_{cw}	Yield of the crop in mixture ($g\ m^{-2}$)	2.2	-
Y_{wc}	Yield of the weeds in mixture ($g\ m^{-2}$)	2.2	-
YL	Yield loss (%)	2.6	-
z	Coefficient in yield loss equation	2.9	-
$\sin\beta$	Sine of solar elevation (-)	4.1	DSINB
γ	Psychrometric constant (mbar °C^{-1})	5.9	PSYCON
ε	Initial light use efficiency of leaf CO_2 assimilation ($kg\ CO_2\ ha^{-1}\ leaf\ h^{-1}\ /\ (J\ m^{-2}\ leaf\ s^{-1})$)	4.20	EFF
λ	Latent heat for vaporization of water ($J\ kg^{-1}$)	5.7	LHVAP
θ	Soil moisture content ($kg\ H_2O\ m^{-2}$ or mm)	-	

θ_a	Actual soil moisture content (kg H_2O m^{-2} or mm)	5.14	VSMT, VSMRTZ
θ_{cr}	Critical soil moisture content (kg H_2O m^{-2} or mm)	5.14	VSMCR
θ_{fc}	Soil moisture content at field capacity (kg H_2O m^{-2} or mm)	5.14	VSMFC
θ_{wp}	Soil moisture content at permanent wilting point (kg H_2O m^{-2} or mm)	5.14	VSMWP
ρ	Reflection coefficient of a canopy with a spherical leaf angle distribution for *PAR* (-)	4.4	REFVS
ρ_g	Refection coefficient for global radiation (albedo, -)	5.6	REFCFC
σ	Scattering coefficient of leaves for visible radiation (-)	4.4	SCV
σ_{sb}	Stefan-Boltzmann constant (J m^{-2} d^{-1} K^{-4})	5.4	STBC

Index